CONTROLLING BIOFOULING IN SEAWATER REVERSE OSMOSIS MEMBRANE SYSTEMS

T0361974

NIRAJAN DHAKAL

Cover page photo credit

http://midesh2020.eu/publications/

CONTROLLING BIOFOULING IN SEAWATER REVERSE OSMOSIS MEMBRANE SYSTEMS

DISSERTATION

Submitted in fulfillment of the requirements of
the Board for Doctorates of Delft University of Technology
and
of the Academic Board of the UNESCO-IHE
Institute for Water Education
for
the Degree of DOCTOR
to be defended in public on
Thursday 30 November 2017 at 15:00 hours
in Delft, the Netherlands

by

Nirajan DHAKAL

Master of Science in Water Supply Engineering
UNESCO-IHE, Institute for Water Education

born in Gorkha, Nepal

This dissertation has been approved by the
Promotor: Prof. dr. M.D. Kennedy and
Copromotor: Dr. ir. S.G. Salinas Rodriguez

Composition of Doctoral Committee:

Chairman	Rector Magnificus TU Delft
Vice-Chairman	Rector IHE-Delft
Prof. dr. M.D. Kennedy	IHE-Delft/TU Delft, promotor
Dr. ir. S.G. Salinas Rodriguez	IHE-Delft, copromotor

Independent members:

Prof. dr. ir. W.G. J. van der Meer	University of Twente
Prof. dr. ir. J. Miguel Veza	Universidad de Las Palmas, Gran Canaria
Prof. dr. ir. L. C. Rietveld	TU Delft
Dr. ir. B. Blankert	Oasen, the Netherlands
Prof. dr. ir. M.E. McClain	TU Delft/IHE-Delft, reserve member

Published by:
CRC Press/Balkema
Schipholweg 107C, 2316 XC, Leiden, the Netherlands
Pub.NL@taylorandfrancis.com
www.crcpress.com – www.taylorandfrancis.com
ISBN 978-0-8153-5718-6

Acknowledgement

This research presented in this thesis was carried out at UNESCO-IHE Institute for Water Education with financial support from Wetsus, European Centre of Excellence for Sustainable Water Technology.

I wish to express my sincere gratitude to my promotor, Prof. Maria D. Kennedy, Prof. Jan C. Schippers and Sergio Salinas Rodríguez for their strong support, guidance and encouragement during my PhD research project. Their critical thinking and probing discussions throughout the project were valuable in shaping the research. It was a privilege to work with you!

I would like to thank all the participants of the Wetsus research theme "Biofouling" for the productive interaction with both academia and the water industry (Ania, Antoine, Arie, Bas, Bastiaan, Bert, Caroline, Charu, Fons, Georg, Hilde, Joop, Koen, Leo, Marcel, Mark, Mieke, Natascha, Paula, Remon, Rik). It has been a learning experience interacting with both academics and practitioners in the biofouling research theme, and this provided an opportunity to see how scientific research is applied in practice. In addition, I am indebted to the laboratory staff at wetsus, especially Mieke Kersaan for all her support and patience with the LC-OCD analyses. I sincerely apologize to those who are not mentioned in this acknowledgment.

This work involved pilot testing in Jacobahaven, the Netherlands. This would never have been possible without the commitment and effective coordination from the R&D group of Pentair X-Flow (Remon Dekker, Leo Vredenbregt, Tom Spanjer, Henry Hamberg, and Sander Brinks). Likewise, the support and commitment from de Zeeschelp B.V. (Marco Dubbeldam, Bernd van Broekhoven, and Hanno), Biaqua (Lute Broens, Sandie Chauveau and Monica Paravidino) and UNESCO-IHE (Fred Kruis) were valuable for the success of the project. Besides, I was lucky to supervise four hard-working master students who directly contributed to this research, namely: Alaa Samir Ouda, Joshua Ampah, Mohamed Ismail Nazeer, and Nizordinah Sithole, who have performed the most challenging experiments in this project.

A doctoral degree is not achieved on one's own. The technical and non-technical support of my colleagues was crucial for the completion of this thesis. My gratitude goes to all friends and colleagues (in alphabetic order): Abdulai Salifu, Ahmed Mahmoud, Almotasembellah Abushaban, Assiyeh Tabatabai, Bianca Wassenaar, Chol Abel, Chris Metzker, Emmanuelle Prest, Ferdi Battes, Ferry Horváth, Fiona Zakarika, Frank Wiegman, Frans Knops, Fred Kruis, Iosif Skoullos, Jolanda Boots, Jeroen Lodeweeg, Lea Tan, Loreen Ople Villacorte, Lute Broens, Lyzette Robbemont, Madapura Eregowda, Mariëlle van Erven, Mariska Ronteltap, Matt Luna, Mohanasundar Radhakrishnan, Mohaned Sousi, Mohaned Abunada, Muhammad Dikman Maheng, Muhammad Nasir Mangal, Peter Heerings, Peter Mawioo, Rinnert Schurer, Shreedhar Maskey, Shrutika Wadgaonkar, Sylvia van Opdorp Stijlen, Taha Al Washali, Vanessa Temminck, Water supply chair group of IHE Delft (Branislave Petrusevski, Giuliana Ferrero, Nemaja Trifunovic, Saroj Sharma and Yness March Slokar) and Yuli Ekowati. I sincerely apologize to those who are not mentioned in this acknowledgment.

I am very grateful to my family in Nepal for their never-ending support, encouragement, and love during my stay abroad. My special gratitude goes to my mother and father who sacrificed a lot in their life. I would not be in this stage without their never-ending support. I will always be very grateful my dear mother who passed away during my Ph.D. research period; you will be in my heart forever. My wife Anita Dhamala and my little daughter Angila Dhakal, this work would have been tough without you. I am so blessed to be surrounded by you.

Last but not the least, I thank God for giving me the opportunity, strength, and determination to complete this Ph.D. thesis.

Nirajan Dhakal,

29 October 2017
Delft, the Netherlands

To my late mother, Usha Kumari Dhakal

Summary

Seawater reverse osmosis (SWRO) is the preferred technology of choice for seawater desalination. However, membrane fouling is a major challenge for the cost-effective operation of membrane based desalination systems. An emerging threat to SWRO is the occurrence of algal blooms and the associated high concentration of algal cells and algal organic matter (AOM) in seawater. To help minimize membrane fouling, SWRO systems are equipped with pre-treatment systems. However, current pre-treatment systems are not capable of removing all AOM from SWRO feed water. The AOM that passes from pre-treatment systems accumulates on the SWRO membrane surfaces and acts as a "*conditioning layer*" and can initiate biofilm development in the presence of available nutrients (C, P) in RO feed water.

One notable example was the severe red tide algal bloom in the Middle East in 2008-2009. During this period pre-treatment processes such as granular media filter (GMF) with coagulation suffered from rapid clogging and produced poor quality water for the downstream SWRO system (SDI >5). As a result, some SWRO desalination plants in the coastal areas of the region were forced to shut down to avoid irreversible fouling of their RO membranes. After this event, the application of low-pressure membranes such as microfiltration and ultra-filtration (MF/UF) have been considered as a more reliable pre-treatment during algal blooms. Previous studies have shown that conventional UF membranes are also not capable of removing all organic matter (AOM) from SWRO feed water, and thus organic/biofouling in downstream SWRO could occur. Therefore, new pre-treatment technologies that can remove AOM, as well as other nutrients (C, P) from SWRO feed water are needed to delay the onset of organic and/or biofouling in SWRO systems. Furthermore, better *methods/tools* are required to assess and improve pre-treatment processes in terms of their ability to reduce re-growth potential prior to SWRO membranes.

The overall goal of this research was to assess the ability of conventional UF (150 kDa) and tight UF (10 kDa) either alone or in combination with phosphate removal technology (PRT™) to delay the onset of organic/biological fouling in SWRO feed water during algal blooms. The three main objectives of the research were; i) to better understand ultrafiltration membrane fouling and the root causes of poor backwashability of organic matter generated by different marine algal species, ii) to develop an improved method to measure bacterial regrowth potential (BRP) prior to SWRO membrane systems, and iii) to apply the improved BRP method at laboratory, pilot and full scale to assess the ability of conventional UF (150 kDa) and tight UF (10 kDa) alone and in combination with a phosphate adsorbent to reduce regrowth potential and delay the onset of organic/biological fouling in SWRO feed water during algal blooms.

The study developed an improved method to measure bacterial regrowth potential (BRP) in seawater samples. During the method development, flow cytometry combined with fluorescence staining (SYBR® Green I and Propidium Iodide) was used with a natural consortium of marine bacteria as inoculum. The Level of Detection (LOD) of the method was lowered by developing a standard protocol to prepare blank seawater. The two aspects considered were i) minimize the level of contamination that might originate from sample bottles, chemicals, pipettes and the laboratory environment during blank seawater preparation and ii) minimize leaching of carbon from filters and all surfaces during BRP measurements.

The limit of detection of BRP method was lowered to $43 \times 10^3 \pm 12 \times 10^3$ cells/mL, which is equivalent to 9.3 ± 2.6 µg-$C_{glucose}$/L assuming a yield factor of 4.6×10^6 cells/µg-C for marine bacteria. Calibration of the method was performed with glucose as a standard substrate in artificial and natural seawater. The BRP method was applied in full-scale seawater desalination plants in the Middle East to assess the biofouling potential of SWRO feed water, as well as to assess the performance of the pre-treatment systems.

The next phase of the study was to investigate the fouling potential and fouling behavior of algae and algal released organic matter in ultrafiltration membranes. For this purpose, four marine algae were cultivated namely: *Chaetoceros affinis (Ch), Rhodomonas balthica (Rb), Tetraselmis suecica (Te),* and *Phaeocystis globulosa (Ph).* During the growth and stationary/decline phase, the algal cell density, chlorophyll-a, biopolymer, transparent exopolymer particles (TEP) concentration and MFI-UF$_{10kDa}$ (membrane fouling potential) were measured. Fouling experiments were executed with capillary ultrafiltration, filtration inside to outside, and backwashable and non- backwashable fouling was monitored.

During the growth, stationary/decline phase of the algal species remarkable differences were observed in the production of biopolymers, TEP and MFI-UF$_{10kDa}$. Membrane fouling potential (MFI-UF$_{10kDa}$) was linearly related to algal cell density and chlorophyll-a concentration, biopolymer concentration, TEP, during the growth phase of the algal species.

After the growth phase, the relationship between MFI-UF$_{10kDa}$ and algal cell density and chlorophyll-a concentration did not continue. In experiments with capillary ultrafiltration, membranes (150 kDa) fed with water having 0.5 mg-biopolymer - C/L back washable fouling coincided with the MFI-UF$_{150kDa}$ and TEP for *Rh, Te, and Ph*. Back washable fouling for *Ch* deviated and was substantially higher. The non-back washable fouling of the ultrafiltration membranes varied strongly with the type of algal species and coincided with MFI-UF$_{150kDa}$ and TEP concentration. *Rh* demonstrated the highest and *Ph* the lowest non-back washable fouling (at a level of 0.5 mg-biopolymer-C/L) in the feed water. This non-backwashable fouling is attributed to polysaccharides (stretching - OH) and sugar ester group (stretching S=O) present in the AOM. Furthermore, the characterization of permeate quality of UF showed biopolymer rejection of 60 % to 80 % depending upon on the algal species. This indicates that biopolymers having a size smaller than the pores of the ultrafiltration membranes may also contribute to non-backwashable fouling in UF/RO systems. Therefore, a more robust pre-treatment is needed with enhanced removal of AOM from RO feed water in order to delay the onset of organic/biological fouling in SWRO systems.

The next phase of the study was to apply the improved BRP method and other analytical tools such as transparent exopolymer particles (TEP), modified fouling index (MFI-UF), liquid chromatography organic carbon detection (LC-OCD) to assess the biofouling reduction potential of tight ultrafiltration (10 kDa) pre-treatment. The tests were performed at laboratory and pilot scale and were performed using MF and UF membranes with a wide range of MWCO and algal organic matter (AOM) produced by *Chaetoceros affinis* as a feed solution. The AOM rejection experiments performed with MF and UF membranes showed 3-4 times lower biopolymer and TEP concentration as well as MFI-UF$_{10kDa}$ with tight UF (10 kDa) compared with the permeates of high MWCO MF and UF membranes. The measured bacterial regrowth potential (BRP) of tight UF permeate was *ca.* 2-3 times lower than the permeate of high MWCO MF and UF membranes. However, it should be noted that no remarkable difference was observed in bacterial regrowth potential of tight UF (10 kDa) and high molecular weight cut off (150 kDa) UF in pilot-scale experiments.

Biofouling experiments performed at pilot-scale using the permeate of tight UF (10 kDa) and conventional UF (150 kDa) showed no substantial head loss development in the membrane fouling simulator (MFS) monitors in short-term (15 days) experiments. The measured biomass accumulated in the MFS monitor fed with 10 kDa UF permeate was *ca.* 860 pg ATP/cm^2, which was 2 and 5 times lower than measured in MFS monitors fed with 150 kDa

UF permeate and UF feed, respectively. In terms of hydraulic operation, the tight UF showed 1.5 times higher non-backwashable fouling rate development compared with a 150 kDa UF. This could be attributed to the lower surface porosity of the 10 kDa UF membrane, which resulted in lower backwashing and chemical enhanced backwashing (CEB) efficiency compared to the 150 kDa UF. Improving the surface porosity of the 10 kDa UF may lower non-backwashable fouling rate development. In general, the results from the laboratory and pilot-scale demonstrated the potential for tight UF (10 kDa) as a pretreatment for SWRO during algal blooms, but validation in long-term experiments is still necessary.

The role of phosphate removal technology (PRT™) combined with tight ultrafiltration (10 kDa) in delaying the onset of biofouling in SWRO systems was also investigated. Laboratory-scale experiments showed that the application of PRT™ resulted in improved removal of biopolymers as well as dissolved phosphate from SWRO feed water, compared with ultrafiltration alone. Furthermore, the application of PRT™ substantially lowered the bacterial regrowth potential (BRP) of UF permeate sample independent of the pore size of the UF membrane. The addition (spiking experiment) of 10 μg PO_4 – P/L to the permeate of UF-PRT™ resulted in a significantly higher (by factor 2) bacterial regrowth potential, suggesting that the removal of phosphate limited bacterial regrowth.

Finally, biofouling experiments using membrane fouling simulator (MFS), showed no increase in feed channel pressure drop in MFS units fed with permeate of tight UF (10 kDa) followed by a phosphate adsorbent (PRT™) for at least 21 days when operated at a cross flow velocity of 0.2 m/s. Moreover, a pressure drop of approximately 500 mbar was observed in MFS units fed with permeate of tight UF (10 kDa) alone when operated for the same period and with similar conditions and thus illustrates the role of the phosphate adsorbent (PRT™) in delaying the occurrence of biofouling.

Membrane autopsies also showed that the biomass accumulation in the MFS fed with permeates of UF+PRT™ was below the detection limit. While the measured ATP was 6,000 pg ATP/cm^2 in the MFS fed with permeate of tight UF alone. The higher biomass accumulation in the MFS fed with the permeate of the tight UF (10 kDa) alone could be attributed to the passage of low molecular weight (LMW) organic and dissolved phosphate through the 10 kDa UF. The possible contribution of LMW organics (tested using EDTA) showed a linear regrowth ($R^2=0.65$) between EDTA concentration and the net bacterial regrowth. Overall, the proof of principle experiments demonstrated that the removal of phosphate by the application of PRT™ combined with UF (10 kDa) restricted biomass growth and may thus delay the onset of biofouling in SWRO membranes. Moreover, a more extended period of testing is needed for further verification of both technologies.

Overall, this study demonstrated that an improved bacterial regrowth potential (BRP) method can be used to (i) assess pre-treatment technology in terms of BRP reduction, (ii) monitor the performance of pre-treatment systems and (iii) develop essential strategies to mitigate membrane fouling in SWRO systems. This study also demonstrated that the removal of algal organic matter (AOM), and dissolved phosphate from SWRO feed water is a potential strategy to delay the onset of organic and biofouling in SWRO systems during algal blooms. Tight UF (10 kDa) coupled with an adsorbent to remove phosphate showed higher potential compared to UF alone (10 kDa) with respect to AOM and nutrient (C, P) removal.

Finally, it is still necessary to further develop existing and new methods that can detect low concentrations of nutrients e.g. carbon and phosphate in seawater, to support the development of improved membrane fouling prevention strategies

Table of Contents

1

General introduction

Contents

1.1 Background

The global demand for water has increased over the past decades due mainly to i) population growth, ii) increase in per capita water demand, iii) expanded irrigation schemes, and iv) economic development (Curmi et al., 2013, de Graaf et al., 2014). Furthermore, uneven rainfall distribution, uneven population distribution, and unequal water use distribution have increased the regional water scarcity. The 2015 UN report on the Millennium Development Goals stated, "Water scarcity affects more than 40 % of the population." As projected by the International Water Management Institute, more than half of the world population will suffer from water scarcity by the year 2025 (Figure 1.1). Thus, it is imperative to locate other water resources such as wastewater, water reuse, and seawater desalination to increase freshwater production in working to alleviate the global water crisis.

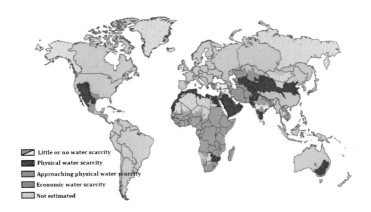

Little or no water scarcity
Physical water scarcity
Approaching physical water scarcity
Economic water scarcity
Not estimated

Figure 1.1: Projected global water scarcity in 2025 (IWMI, 2006)

Freshwater comprises about 2.5 % of the total amount of water on the planet, and the rest (97.5 %) is salt water. The available freshwater is not evenly distributed around the world, with variations over geographical regions and time. Only a small part of available freshwater resources is a naturally renewable source of freshwater (Miller, 2003).

Today, desalination is one of the solutions that are increasingly applied to solve freshwater scarcity problems in many regions of the world. Although desalination is the best known to produce freshwater from seawater, it can also be used to treat slightly salty (brackish) water, low-grade surface, and groundwater. Of the various desalination technologies, reverse osmosis (RO) is the most widely used desalination system for brackish and seawater. The expansion of RO globally has been relatively rapid since the year 2000 and is expected to reach the cumulative capacity of about 60 Mm³/day by the year 2018 (Figure 1.2). Almost half (47 %) of RO-desalinated water is from seawater, and the rest is mainly from brackish, freshwater and treated wastewater (DesalData, 2016).

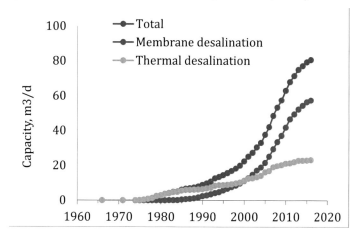

Figure 1.2: Seawater reverse osmosis plant capacity from 1970-2018 (DesalData, 2016)

Membrane-based seawater desalination is currently dominating the market mainly because of its reduction in the power consumption and per unit production cost. As illustrated in Figure 1.3 b, the power consumption reduced from 16 kWh/m³ in 1970 to 2 kWh/m³ in 2008. Likewise, the cost per cubic meter has also decreased from $ 1.6/m³ to $ 0.6/m³ from 1982 to 2010 (Figure 1.3 a).

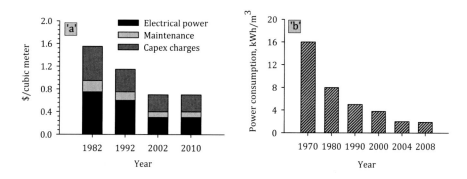

Figure 1.3: Trends of a) cost in $/m³ (WaterReuseAssociation, 2012) and b) power consumption in kWh/m³ (Elimelech, 2012) in Seawater reverse osmosis plants

1.2 Algal blooms and seawater reverse osmosis operation

Many seawater reverse osmosis (SWRO) desalination plants are located in and treat seawater from coastal zones where algal blooms frequently occur (Caron et al., 2010). Even as SWRO plants are rapidly increasing throughout the world, their operations is affected by the occurrence of algal blooms, which cause problems including membrane fouling. One example is the severe "red tide" bloom that occurred in the Middle East Gulf region (2008 - 2009), caused by the dinoflagellate *Cochlodinium polykrikoides* (Richlen et al., 2010). This bloom forced the stoppage of operations in at least five seawater desalination plants located in that region mainly due to: i) clogging of granular media filters and ii) higher silt density index (SDI >5) in the RO feed water (Pankratz, 2008, Reddy, 2009). This incident exposed the vulnerability of seawater RO plants during severe algal blooms situations. Therefore, efficient removal of algae and algal organic matter (AOM) by pretreatment systems is crucial to minimize such operational problem in SWRO systems.

Algal blooms are unpredictable events and can last from a few days to several months, depending on the life cycles of causative species, environmental conditions and nutrient availability (Villacorte, 2014). Some of the common bloom-forming algal species are illustrated in Figure 1.4. The size of which ranges from 2 μm to 2 mm, and the cell concentration range from 1,000 to 600,000 cells/mL (Villacorte et al., 2015b).

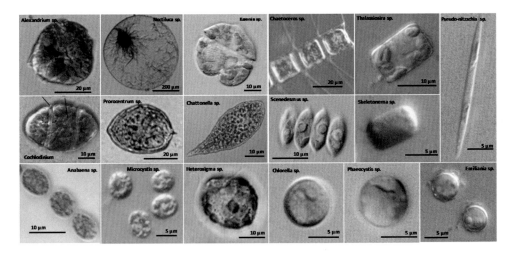

Figure 1.4: Common species of bloom-forming algae in fresh and marine environments (adapted from Villacorte et al., 2014)

The two most important current indicators for monitoring the occurrence of algal blooms are algal cell concentration (>1million cells/L) and chlorophyll-a concentration (>10μg/L)(http://www.waterman.hku.hk, 2016). Figure 1.5 shows the level of average chlorophyll-a concentration, measured in 2009, in surface water bodies worldwide. The red color in the map shows the chlorophyll-a concentration > 10μg/L measured in all coastal regions. This illustrates that desalination plants located in coastal zones are vulnerable to algal blooms.

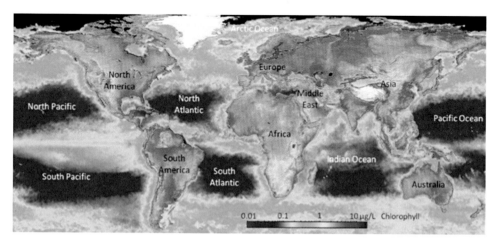

Figure 1.5: Typical average annual distribution of chlorophyll-a in surface water bodies on Earth (adapted from Villacorte et al., 2014)

Algal blooms increase the suspended solids concentration as well as the organic substances responsible for membrane fouling (Caron et al., 2010). During algal blooms, algae release algal organic matter (AOM) which has been shown to be the leading cause of membrane fouling, rather than the algae themselves (Ladner et al., 2010, Qu et al., 2012, Schurer et al., 2012, Villacorte et al., 2015a). The AOM mainly consists of polysaccharides, proteins, lipids, nucleic acids, and other dissolved organic substances (Fogg, 1983, Myklestad, 1995). A significant fraction of AOM is the transparent exopolymer particles (TEP) (Villacorte et al., 2013) which are highly sticky polysaccharides and glycoproteins (Passow et al., 1995). The presence of TEP-like materials thus causes or initiates organic fouling in ultrafiltration (UF) and biological fouling in UF/RO membranes (Berman et al., 2005, Berman et al., 2011, Kennedy et al., 2009). The consequences of membrane fouling in reverse osmosis lead to:

- Increase in head loss across the feed spacer of spiral wound elements

- Higher energy consumption to maintain the constant flux operation
- Higher chemical cleaning frequency
- Increase in the replacement of membranes due to irreversible membrane fouling
- Decrease in the rate of water production due to longer downtime during chemical cleaning and membrane replacement
- Increase in salt passage and a resulting deterioration of permeate quality

To help prevent membrane fouling, SWRO plants install pre-treatment systems (e.g., media filters with coagulation or MF/UF). The particulate and colloidal fouling in SWRO is mostly controlled with such existing pre-treatment; however, the occurrences of organic and biological fouling are still a significant issue in SWRO membranes.

1.3 Pre-treatment for seawater reverse osmosis

Seawater reverses osmosis plants are equipped with pre-treatment systems to ensure consistent performance of the SWRO membranes, and to reduce cleaning frequency. Pre-treatment consists of the intake and screening systems, processes for particulate matter removal and control of biological growth (Huehmer et al., 2006). However, inadequate pre-treatment is still the leading cause of SWRO system failure (Gallego et al., 2007). Pre-treatment for SWRO systems can be i) conventional or ii) advanced or iii) a combination of both, depending upon the raw water quality. Conventional treatment systems typically consist of coagulation and (dual) media filtration using a variety of filtration media such as combinations of single and two-stage, gravity, and pressurized media filters. However, media filters demonstrated operational problem during the 2008 - 2009 algal blooms in the Gulf of Oman. The associated problems were clogging of media filters and poor effluent water quality (SDI > 5), which forced SWRO plants to shut down (Richlen et al., 2010). Furthermore, these systems are also characterized by high coagulant consumption.

Dissolved air flotation (DAF) has recently gained attention as a promising pre-treatment option during algal blooms. DAF is a clarification process that is typically applied before media filters and MF/UF with the aim of removing particles (Cleveland et al., 2002). It has been reported that the algal cell removal efficiency of DAF is better (90 – 99 %) than sedimentation (60 – 90 %) (Gregory and Edzwald (2010), cited by (Villacorte, 2014). However, a high coagulant dose of up to 20 mg/L as FeCl$_3$ is needed to achieve adequate removal of algae by DAF units (Rovel, 2003).

Recently, application of low-pressure membrane systems (MF/UF) is increasing as pre-treatment to treat seawater during algal blooms (Villacorte et al., 2015a, Voutchkov, 2010). MF/UF pre-treatment has numerous advantages compared to conventional pre-treatment: mainly lower footprint, higher permeate quality, higher rejection of organics, and lower chemical consumption (Pearce, 2007, Wilf et al., 2001). Furthermore, UF operated with inline coagulant dosing, with low concentration, during algal blooms has demonstrated stable hydraulic operation (Schurer et al., 2013).

Moreover, several studies have shown that existing pre-treatment systems are often useful in removing algae itself. However, the systems allow the passage 30 – 80 % of algal released organic matter (AOM), measured as biopolymer concentration as shown in Figure 1.6 (Guastalli et al., 2013, Salinas - Rodriguez et al., 2009, Tabatabai et al., 2014). The wide range of removal efficiency of pre-treatment depends on the amount of coagulant applied during operation. As illustrated in Figure 1.6, pre-treatment by beach wells showed the highest biopolymer rejection (> 80 %) compared to other pretreatment processes. However, beach wells are not feasible for large desalination plants.

Overall, the passage of 30 – 80 % of algal biopolymer from the existing pre-treatment systems may accumulate on downstream SWRO membranes. The deposited biopolymers on a SWRO membrane may act as a conditioning layer, where bacteria can grow and multiply to form a biofilm in the presence of available nutrients from feed water. Therefore, SWRO operations with current pre-treatment systems are still vulnerable mainly due to possible organic and biological fouling during algal blooms.

Despite the fact that existing pre-treatment technologies cannot protect RO operation during algal blooms, an increasing number of large-scale RO plants (> 500,000 m³/day) will be installed in the coming years (Kurihara et al., 2013). Such large-scale plants may be threatened by algal blooms and thus demand a robust pretreatment technology to minimize problems of organic/biofouling occurrence in SWRO membranes.

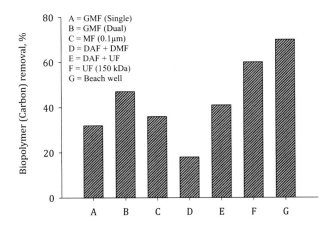

Figure 1.6: Biopolymer removal by various SWRO pre-treatment systems tested with various water source i) Western Mediterranean Sea (B, C, G), ii) with Eastern Mediterranean Sea (A) (Salinas - Rodriguez et al., 2009), iii) Western Mediterranean Sea (D, E) (Guastalli et al., 2013) and iv) laboratory cultured AOM from "Chaetoceros affinis" (F) (Alizadeh Tabatabai et al., 2014).

1.4   Future generation of pre-treatment in SWRO

Membrane fouling (organic and biological) remains a major limitation of SWRO desalination operation, despite improvements in pre-treatment systems. A new generation of pre-treatment technology is needed to further protect the performance of SWRO operation, showing better removal of algal biopolymers as well as the ability to limit the concentration of nutrients (C, P) so that downstream SWRO will not suffer from organic and biofouling. Removal of biopolymers from the feed water in itself can substantially delay biofouling, as there is no *"conditioning layer"* to initiate biofilm development. On the other hand, limiting essential nutrients from the SWRO feed water may delay the onset of biological growth in the system.

The application of tight ultrafiltration (10 kDa) may eliminate fouling caused by algal-derived biopolymers in RO systems. Although tight UF is expected to remove particulate and colloidal material from the feed water efficiently, it is not capable of removing dissolved nutrients such as carbon and phosphate as UF membranes are not designed to remove dissolved carbon and phosphate. However, the removal of such dissolved nutrients from SWRO feed water might contribute to delaying onset of biofouling in SWRO systems.

It has been demonstrated that limiting phosphate in SWRO feed water can control biofouling (Jacobson et al., 2009). It was also reported that phosphate limitation can prevent the occurrence of biofouling in RO systems even in the presence of high concentrations of other nutrients (Vrouwenvelder et al., 2010). Various technologies or methods are available that can remove phosphate from water but are reported not sustainable (Sevcenco et al., 2015). This provided an opportunity to look for alternative phosphate removal technology or methods which will are sustainable and environmentally friendly.

In this study, the application of pre-treatment systems that can remove substantial amounts of AOM and nutrients such as carbon and phosphate and eventually delay the onset of organic and biofouling in SWRO systems were investigated. The potential of tight UF and a newly developed phosphate adsorbent were tested to verify the following hypotheses;

- Tight ultrafiltration (10 kDa) having a lower molecular weight cut off than the conventional UF (150 kDa) is expected to be more effective in removing organic matter and thus delay the onset of biofouling in SWRO

- A phosphate adsorbent is capable of reducing the phosphate in SWRO feed water to such a level that it can limit bacterial regrowth.

It is expected that the integration of the two technologies as pre-treatment can be a promising solution to control organic and biofouling in SWRO systems as illustrated in Figure 1.7.

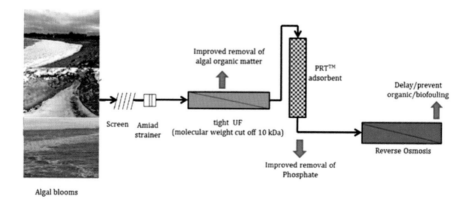

Figure 1.7: Research concept

1.5   Goal and objectives

The overall aim of this study was to assess the ability of conventional and tight UF either alone or in combination with phosphate removal technology (PRT™) to delay the onset of organic/biological fouling in SWRO feed water during algal blooms.

This specific objectives are the following:

- To understanding ultrafiltration membrane fouling and the root causes of poor backwashability of organic matter generated by four different marine algal species.
- To develop an improved method to measure bacterial regrowth potential (BRP) prior to SWRO membrane systems
- To apply the improved BRP method at laboratory, pilot and full scale to assess the ability of conventional UF (150 kDa) and tight UF (10 kDa) alone and in combination with a phosphate adsorbent to reduce regrowth potential and delay the onset of organic/biological fouling in SWRO feed water during algal blooms.

1.6   Outline of the thesis

This thesis has been structured into seven chapters as described below:

Chapter 1 is a general introduction on the background of the study, research problems, and the needs of the future generation of advanced pre-treatment systems to eliminate or delay the onset of biofouling in seawater reverse osmosis during algal blooms. This chapter also includes research concept, goal, and objectives of the study.

Chapter 2 is a review of the perspective and challenges for the global desalination market.

Chapter 3 describes the development of improved method to measure bacterial regrowth potential (BRP) in seawater using a natural bacterial consortium as inoculum in combination with flow cytometry.

Chapter 4 describes the fouling of ultrafiltration membranes by organic matter generated by four marine algal species

Chapter 5 describes the role of tight ultrafiltration (UF) (with a molecular weight cut off of 10 kDa) in reducing the biofouling potential of SWRO feed water during algal blooms. The proof of principle was performed at laboratory and pilot scales using various pore size MF/UF membranes.

Chapter 6 describes the role of phosphate removal technology (PRT™) combined with tight UF in reducing biofouling potential of SWRO feed water during algal blooms. The proof of principle was performed at laboratory and pilot scale.

Chapter 7 provides a summary of conclusions and outlook.

1.7 References

Alizadeh Tabatabai, S.A., Schippers, J.C. and Kennedy, M.D. (2014) Effect of coagulation on fouling potential and removal of algal organic matter in ultrafiltration pretreatment to seawater reverse osmosis. Water Research 59, 283-294.

Berman, T. and Holenberg, M. (2005) Don't fall foul of biofilm through high TEP levels. Filtration & Separation 42(4), 30-32.

Berman, T., Mizrahi, R. and Dosoretz, C.G. (2011) Transparent exopolymer particles (TEP): A critical factor in aquatic biofilm initiation and fouling on filtration membranes. Desalination 276(1–3), 184-190.

Caron, D.A., Garneau, M.-È., Seubert, E., Howard, M.D.A., Darjany, L., Schnetzer, A., Cetinić, I., Filteau, G., Lauri, P., Jones, B. and Trussell, S. (2010) Harmful algae and their potential impacts on desalination operations off southern California. Water Research 44(2), 385-416.

Cleveland, C., Hugaboom, D., Raczko, B. and Moughamian, W. (2002) DAF pretreatment for ultrafiltration: cost and water quality implications.

Curmi, E., Richards, K., Fenner, R., Allwood, J.M., Kopec, G.M. and Bajželj, B. (2013) An integrated representation of the services provided by global water resources. Journal of Environmental Management 129(0), 456-462.

de Graaf, I.E.M., van Beek, L.P.H., Wada, Y. and Bierkens, M.F.P. (2014) Dynamic attribution of global water demand to surface water and groundwater resources: Effects of abstractions and return flows on river discharges. Advances in Water Resources 64(0), 21-33.

DesalData (2016) Worldwide desalination inventory (MS Excel format), Available from www. DesalData.com on June 2016.

Elimelech, M. (2012) Seawater Desalination, 2012 NWRI Clarke prize conference, Newport Beach, California.

Fogg, G. (1983) The ecological significance of extracellular products of phytoplankton photosynthesis. Botanica Marina 26 (1), 1-43.

Gallego, S. and Darton, E. (2007) Simple laboratory techniques improve the operation of RO pre-treatment systems, Maspalomas, Gran Canaria.

Guastalli, A.R., Simon, F.X., Penru, Y., de Kerchove, A., Llorens, J. and Baig, S. (2013) Comparison of DMF and UF pre-treatments for particulate material and dissolved organic matter removal in SWRO desalination. Desalination 322, 144-150.

http://www.waterman.hku.hk (2016) Assessing the occurrence of an algal bloom - Chlorophyll-a concentration.

Huehmer, R. and Henthorne, L. (2006) Advance in RO pretreatment techniques, Haifa, Israel.

Jacobson, J.D., Kennedy, M.D., Amy, G. and Schippers, J.C. (2009) Phosphate limitation in reverse osmosis: An option to control biofouling? Desalination and Water Treatment 5, 198-206.

Kennedy, M.D., Muñoz - Tobar, F.P., Amy, G.L. and Schippers, J.C. (2009) Transparent exopolymer particles (TEP) fouling of ultrafiltration membrane systems. Desalination and Water Treatment 6 (1-3), 169 - 176.

Kurihara, M. and Hanakawa, M. (2013) Mega-ton Water System: Japanese national research and development project on seawater desalination and wastewater reclamation. Desalination 308(0), 131-137.

Ladner, D.A., Vardon, D.R. and Clark, M.M. (2010) Effects of shear on microfiltration and ultrafiltration fouling by marine bloom-forming algae. Journal of Membrane Science 356(1–2), 33-43.

Myklestad, S.M. (1995) Release of extracellular products by phytoplankton with special emphasis on polysaccharides. The Science of The Total Environment 165(1–3), 155-164.

Pankratz, T. (2008) Red tides close desal plants. Water Desalination Report 44 (1).

Passow, U. and Alldredge, A.L. (1995) A dye-binding assay for the spectrophotometric measurement of transparent exopolymer particles (TEP). Limnology Oceanography 40(7), 1326-1335.

Pearce, G.K. (2007) The case for UF/MF pretreatment to RO in seawater applications. Desalination 203(1–3), 286-295.

Qu, F., Liang, H., Tian, J., Yu, H., Chen, Z. and Li, G. (2012) Ultrafiltration (UF) membrane fouling caused by cyanobacteria: Fouling effects of cells and extracellular organics matter (EOM). Desalination 293, 30-37.

Reddy, V. (2009) Red Tide in the Arabian Gulf. MEDRC Watermark 40(3).

Richlen, M.L., Morton, S.L., Jamali, E.A., Rajan, A. and Anderson, D.M. (2010) The catastrophic 2008–2009 red tide in the Arabian Gulf region, with observations on the identification and phylogeny of the fish-killing dinoflagellate Cochlodinium polykrikoides. Harmful Algae 9(2), 163-172.

Rovel, J.M. (2003) Why a SWRO in Taweelah-pilot plant results demonstrating feasibility and performance of SWRO on Gulf water? In: Proceedings of International Desalination Association World Congress, Nassau, Bahamas.

Salinas - Rodriguez, S.G., Kennedy, M.D., Schippers, J.C. and Amy, G.L. (2009) Organic foulants in estuarine and bay sources for seawater reverse osmosis - Comparing pre-treatment processes with respect to foulant reductions. Desalination and Water Treatment 9, 155-164.

Schurer, R., Janssen, A., Villacorte, L.O. and Kennedy, M.D. (2012) Performance of ultrafiltration & coagulation in a UF-RO seawater desalination demonstration plant. Desalination and Water Treatment 42(1-3), 57-64.

Schurer, R., Tabatabai, A., Villacorte, L., Schippers, J.C. and Kennedy, M.D. (2013) Three years operational experience with ultrafiltration as SWRO pre-treatment during an algal bloom. Desalination and Water Treatment 51(4-6), 1034-1042.

Sevcenco, A.-M., Paravidino, M., Vrouwenvelder, J.S., Wolterbeek, H.T., van Loosdrecht, M.C.M. and Hagen, W.R. (2015) Phosphate and arsenate removal efficiency by thermostable ferritin enzyme from Pyrococcus furiosus using radioisotopes. Water Research 76, 181-186.

Tabatabai, S.A.A., Schippers, J.C. and Kennedy, M.D. (2014) Effect of coagulation on fouling potential and removal of algal organic matter in ultrafiltration pretreatment to seawater reverse osmosis. Water Research 59, 283-294.

Villacorte, L.O. (2014) Algal blooms and membrane - based desalination technology, Ph.D. thesis, UNESCO-IHE.

Villacorte, L.O., Ekowati, Y., Winters, H., Amy, G., Schippers, J.C. and Kennedy, M.D. (2015a) MF/UF rejection and fouling potential of algal organic matter from bloom-forming marine and freshwater algae. Desalination 367, 1-10.

Villacorte, L.O., Ekowati, Y., Winters, H., Amy, G., Schippers, J.C. and Kennedy, M.D. (2013) Characterisation of transparent exopolymer particles (TEP) produced during algal bloom: a membrane treatment perspective. Desalination and Water Treatment 51((4-6)), 1021-1033.

Villacorte, L.O., Tabatabai, S.A.A., Dhakal, N., Amy, G., Schippers, J.C. and Kennedy, M.D. (2015b) Algal blooms: an emerging threat to seawater reverse osmosis desalination. Desalination and Water Treatment 55(10), 2601-2611.

Voutchkov, N. (2010) Considerations for selection of seawater filtration pretreatment system. Desalination 261(3), 354-364.

Vrouwenvelder, J.S., Beyer, F., Dahmani, K., Hasan, N., Galjaard, G., Kruithof, J.C. and Van Loosdrecht, M.C.M. (2010) Phosphate limitation to control biofouling. Water Research 44(11), 3454-3466.

WaterReuseAssociation (2012) Seawater desalination costs, white paper.

Wilf, M. and Schierach, M.K. (2001) Improved performance and cost reduction of RO seawater systems using UF pretreatment. Desalination 135(1–3), 61-68.

2

Perspectives and challenges for desalination

Contents

This chapter is based on the updated version of the paper:

Dhakal, N., Salinas Rodriguez, S.G., Schippers, J.C. and Kennedy, M.D. (2014), Perspectives and challenges for desalination in developing countries, IDA Journal of Desalination and Water Reuse, doi: 10.1179/2051645214Y.0000000015

Abstract

A rapid population growth and urbanization are two main drivers for over-abstraction of conventional freshwater resources in various parts of the world, which leads to the situation of water scarcity (per capita availability < 1,000 m³/year). The projection showed that by 2050, 44 countries (2 billion people) would likely suffer from water scarcity, of which 95 % may live in developing countries. Among them, the countries that would strongly hit by water scarcity by 2050 are Uganda, Burundi, Nigeria, Somalia, Malawi, Eritrea, Ethiopia, Haiti, Tanzania, Niger, Zimbabwe, Afghanistan, Sudan, and Pakistan. Currently, these countries have not yet installed desalination to meet their freshwater demand. However, the current global trend showed that the desalination technology is finding new outlets as an alternative source for supplying water to meet growing water demand in most of the water-scarce countries. The projection showed that these countries would demand desalination capacity of 57 Mm³/day by 2050 to meet the standard of current water demand and to compensate the withdrawal of renewable resources. Case studies from India, China, and South Africa have highlighted that other countries may apply the strategy of using desalinated water to industrial users. Moreover, challenges to the widespread adoption of desalination exist such as expense, significant energy use, the need for specialized staff training, the large footprint of facilities, environmental issues such as greenhouse gas emission (GHGs), chemical discharge and operational problem such as membrane fouling.

Keywords: *Water scarcity, population growth, desalination, developing countries*

2.1   Current trends in a global desalination industry

The large-scale seawater desalination started in the 1960's using thermal distillation processes such as multi-stage flash (MSF) and multi-effect distillation (MED), which dominate the market until 2000 (Figure 2.1). The membrane-based technology (reverse osmosis) was introduced in the market in around 1970's mainly to treat brackish water. Further advancement in the technology and materials made it possible for using RO technology for seawater application since 1980's (Wilf et al., 2007). The desalination market is currently dominated by membrane-based technologies (RO, ED, and NF) since 2000 (Figure 2.1). The average growth in desalination capacity is about 10 % per year in the world, of which membrane desalination consists of 2/3 of the total installed capacity (Dhakal et al., 2014). The total desalination capacity (installed and projected 2018) is about 80 Mm^3/d, of which 75 % ($\sim$ 60 Mm^3/d) is using reverse osmosis technology. In fact, the ratio will likely to change since most of new contracted desalination plants are based on membrane-based technology (DesalData, 2016). Based on the available data, RO is currently the dominant desalination technology and is widely applied for both drinking water and industrial water production. Almost half (47 %) of the RO desalinated water was from seawater and rest mainly from brackish, freshwater and treated wastewater. The extra-large SWRO plants (> 50,000 m^3/d), are already in service, which consists of approximately 58 % of the total installed capacity. The remaining plants (24 %) are categorized as large plants (10,000 - 50,000 m^3/d), 15 % as medium plants (1000 -10,000 m^3/d) and 3 % as small plants (< 1000 m^3/d) as shown in Figure 2.2. As illustrated in figure 2.3, the most of the extra-large plants are located in the Middle East and East/Asia Pacific, Western Europe regions (Figure 2.3).

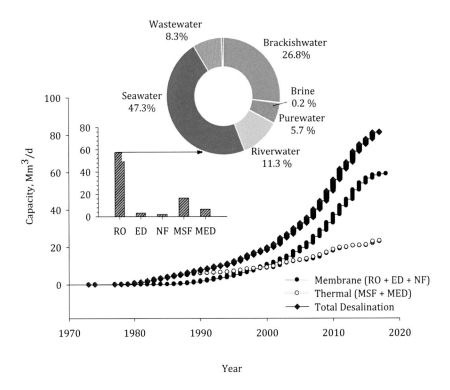

Figure 2.1- Global desalination capacity with regards to desalination technology and RO source water (insert chart) (DesalData, 2016)

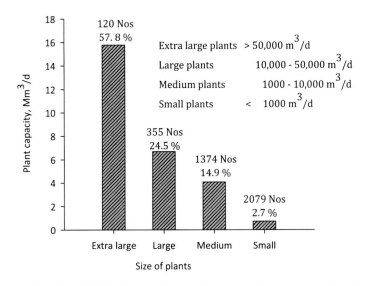

Figure 2.2: Classification of SWRO desalination plants based on their capacity (DesalData, 2016)

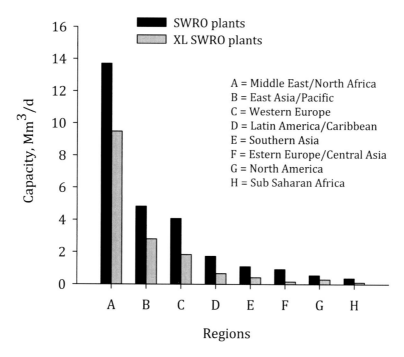

Figure 2.3: Total capacity of SWRO and share of extra-large plants in different regions of the world (DesalData, 2016)

Figure 2.4 illustrates the currently installed and planned SWRO desalination plants (red dots) worldwide, which showed that a high concentration of SWRO plants had been installed in the Middle East, USA, Australia, China, Central Europe, Mediterranean area and Japan. As indicated in the map, most of the desalination plants are located in the coastal line from where SWRO plants abstract raw water, where algal blooms frequently occur (Caron et al., 2010).

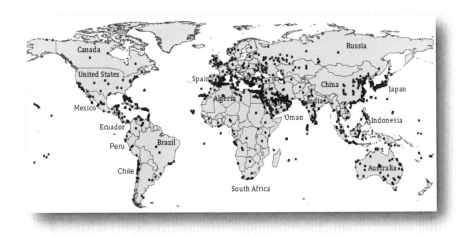

Figure 2.4: Global distribution SWRO plants (red dots). Map processed using ArcGIS 9 and plant coordinates from Desal Data, 2016 (DesalData, 2016)

The membrane-based seawater desalination dominates the market because of reduction in the energy consumption and the cost of operation (Figure 2.5). As illustrated, the energy consumption reduced from 16 kWh/m³ in 1970 to 1.9 kWh/m³ in 2008 (Figure 2.5b). Likewise, the cost needed for electrical power, maintenance, and CapEx charges decreased from $ 1.6/m³ in 1982 to $ 0.6/m³ in 2010 (Figure 2.5 a).

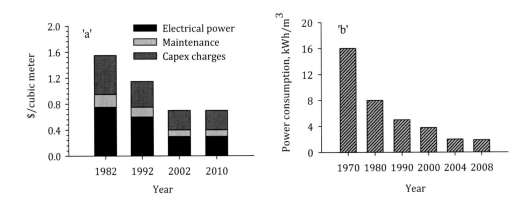

Figure 2.5: Trends of a) cost in $/m³ (Water reuse association, 2012) and b) power consumption in kWh/m³ (Elimelech, 2012) in Seawater reverse osmosis plants

2.2   Is there a need of desalination in developing countries?

The concern over global water availability and its impacts have been uttered during the last decades under the alarming terms of "global water crisis" and water scarcity (Fragkou et al., 2016). The economic and demographic growths are two main drivers for over-abstraction of conventional freshwater resources in various parts of the world (Villacorte et al., 2015), which leads to the situation of water scarcity. Water scarcity is normally considered when the total annual runoff available for human use is less than 1000 m^3/capita/year (Brown et al., 2011). As of 2015, 28 countries mainly in developing countries are suffering from water scarcity. The situation of water scarcity is expected to worsen, as by 2050 the population worldwide is anticipated to reach 9 billion. It has been estimated that by 2050, about 44 countries with a total population of approximately 2 billion would likely suffer from water scarcity (Dhakal et al., 2014), of which, 95 % (1.9 billion) may live in developing countries. The majority of these countries are in Africa and Asia namely, Malawi, Ethiopia, Sudan, Somalia, Nigeria, Uganda, Tanzania, Niger, Zimbabwe, Eritrea, Haiti, Burundi, Pakistan, and Afghanistan (Figure 2.6). The rapid increase in the population growth and the trend of rural-urban migration will intensify the issue of water shortage in these countries mainly due to the withdrawal of fresh water to satisfy the demand for municipal and agricultural use (Bremere et al., 2001). The current available renewable resources in these countries are > 1,000 m^3/cap/year, which will be drastically reduced to below 1,000 m^3/cap/year by 2050 due to the expected population growth (Figure 2.6). The estimation was based on the assumptions that there will be no withdrawal of freshwater resources to fulfill the demand of the increased population. During the projection, the total available renewable water resources which refer to the sum of actual groundwater and surface water in each country was adopted from FAO database (FAOAquastat., 2017) and the total populations (2015 and 2050) was adopted from the World Bank database (WorldBank, 2017).

The potential technical solutions to solve water scarcity are;

Saving water:	Increasing productivity in agriculture & industry
	Reducing leakages in public water supply
	Progressive tariffs
Water transport:	Large distances

Aquifer storage: River water during high flow

Water reuse: Increasing reuse/recycling in industry

                                     & domestic wastewater in agriculture

Desalination   Brackish water, Wastewater, Seawater

Among the different alternative solution to solve the issues of water scarcity, desalination is usually only implemented as a last resort where conventional freshwater resources have been stretched to the limit. Desalination is considered as a drought-proof water source, which does not depend on river flows, reservoir levels or climate change. Desalination may be an option to alleviate scarcity in the industry and coastal cities. The report published by United Nations showed that approximately 44 % of the global population and 8 out of the 10 largest metropolitan area in the world are located within a distance of 150km from the coastline. The rate of population growth in the coastal regions is accelerating, and increasing tourism adds to pressure on the environment (UN Atlas of the Ocean, 2017). Therefore, the possibility of widespread application of seawater desalination in the future is very likely (Villacorte et al., 2015). Although the most well-known application of desalination (and related membrane technology) is to produce freshwater from seawater, it can also be used to treat slightly salty (brackish) water, low-grade surface, and groundwater, and treated effluent resources (Dhakal et al., 2014). The current global trend showed that the desalination technology is finding new outlets as an alternative source for supplying water to meet growing water demand in most of the water-scarce countries (Bremere et al., 2001). However, there have been barriers to its widespread adoption of technology mainly due to its cost, energy, lack of expertise, and the footprint.

2.3   Current and future status of desalination market in 13 water scare countries

The 13 countries that will be strongly hit by water scarcity by 2050 are Uganda, Burundi, Nigeria, Somalia, Malawi, Eritrea, Ethiopia, Haiti, Tanzania, Niger, Zimbabwe, Afghanistan, Sudan, and Pakistan (Figure 2.6). The current desalination status and the potential future market in these water-scarce countries were studied. The current (2016) status of installed seawater and brackish water desalination plants (online, construction and presumed online) and its share for municipal, domestic purpose in each of the selected water-scarce countries is presented in Table 2.1.

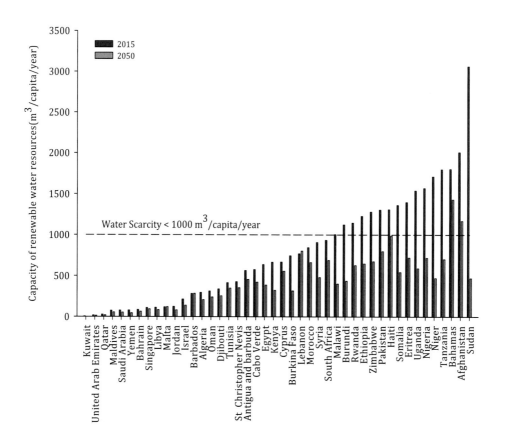

Figure 2.6: Countries expected to be water-scarce by 2050

As illustrated in Table 2.1, countries such as Burundi, Malawi, Niger, Somalia, Uganda, and Zimbabwe have not yet installed any desalination plants. While, other countries such as Afghanistan, Eritrea, Ethiopia, Nigeria, Pakistan, Sudan, and Tanzania have already installed either seawater or brackish water or both desalination plants. The user category in most of the countries that have already installed desalination plants was municipal water and industry.

Table 2.1:Currently installed desalination capacity (sea and brackish water) in the chosen water-scarce countries and its share of municipal, domestic supply (DesalData, 2016)

| Country | Region | Desalination capacity, Q_{2016} | | | |
| | | Seawater | | Brackish water | |
		Capacity, Q_{SW} [m³/day x1000]	Municipal water, Y_{SW}	Capacity, Q_{BW}, [m³/day x 1000]	Municipal water, Y_{BW}
Afghanistan	Central Asia	0.0	0	2.5	0.85
Burundi	East Africa	0.0	0	0.0	0
Eritrea	Northeast Africa	1.0	1.0	0.15	0
Ethiopia	Northeast Africa	1.7	0.42	0.02	1.0
Malawi	Southeast Africa	0.0	0	0.0	0
Niger	West Africa	0.0	0	0.0	0
Nigeria	West Africa	10.9	0.55	4.7	0.15
Pakistan	South Asia	44.6	0.35	93.1	0.02
Somalia	East Africa	0.1	1.00	0.0	0
Sudan	North Africa	43.6	0.81	0.48	0
Tanzania	East Africa	0.6	1.0	6.1	0
Uganda	East Africa	0.0	0	0	0
Zimbabwe	Southern Africa	0.0	0	0	0

The current freshwater withdrawals in these 13 countries were studied based on the available data from FAO database. The general trend showed that most of the water-scarce countries withdraw freshwater mainly for the agricultural activities, municipal use, and industrial use (Table 2.2).

Table 2.2: Water withdrawal in each water-scarce countries (FAOAquastat., 2017)

| Countries | N_{2015} [millions] | Urban population [millions] | Water withdrawal, m³/capita/day | | | |
			Agriculture	Municipal	Industries	Total
Afghanistan	32,5	4,3	0,1	0,1	0,0	0,2
Burundi	11,2	1,2	0,5	0,1	0,0	0,7
Eritrea	5,2	1,0	0,3	0,4	0,1	0,8
Ethiopia	99,4	16,9	0,8	0,1	0,0	1,0
Malawi	17,2	2,8	1,5	0,1	0,0	1,6
Niger	19,9	7,4	1,5	0,1	0,0	1,6
Nigeria	182	38,4	0,0	0,0	0,0	0,0
Pakistan	189	74,5	0,2	0,1	0,0	0,4
Somalia	10,8	2,8	1,3	0,2	0,1	1,6
Sudan	40,2	15,2	3,2	0,0	0,0	3,2
Tanzania	53,5	19,0	12,9	0,1	0,1	13,1
Uganda	39	8,7	4,7	0,2	0,0	4,9
Zimbabwe	15,6	6,1	6,3	0,4	0,1	6,7
Average			2,6	0,146	0,041	2,8

The highest use was in agriculture, which ranged from 0.1 to 12.9 m³/cap/day with an average of 2.6 m³/cap/day. The average withdrawal for the municipal purpose was 0.146 m³/cap/day, which is 178 times lower than in the agricultural sector. The current average per capita municipal, domestic water use, WW_{AVG} = 0.146 m³/cap/d from Table 2.2, was calculated from the municipal water withdrawals in each country and distributed over the urban population in that country. We consider urban population as a potential user of the desalination in future. Based on this the need for the desalination capacity by 2050 in these countries was projected. The following assumptions were made during the projection.

- No withdrawal of renewable water resources to meet the water demand by population growth
- The water demand needed will only be supplied by desalination
- The populations of urban areas are only assumed a potential user of desalinated water.
- The current average withdrawal for the municipal purpose, i.e., 0.146 m³/cap/day is assumed to be constant throughout the projection period

The potential desalination (seawater and brackish water) growth in each of the selected water-scarce countries was calculated using the difference between the desalination capacity (Q_{2050}) and the currently installed desalination capacity (Q_{2016}) using Equation. 2.1 and 2.2 (Bremere et al., 2001)

$$\Delta Q_{2050} = Q_{2050} - Q_{2016} = (N_{2050}. U. WW_{2050}) - (Q_{SW}Y_{SW} + Q_{BW}Y_{BW}) \qquad \text{Equation 2.1}$$

$$Q_{2016} = \frac{Q_{SW}Y_{SW} + Q_{BW}Y_{BW}}{N_{2016} U} \qquad \text{Equation 2.2}$$

Where,

N_{2050} =projected population of each selected water scarce country by 2050

N_{20015} =current population in each country

U   =population share that lives in urban centers

WW_{2050} =per capita municipal, domestic water use by 2050 in m³/cap/d

Q_{SW} and Q_{BW} =currently installed sea and brackish water desalination capacity, m³/d

Y_{SW} and Y_{BW} =share of capacity used for municipal water production.

The projected growth in the desalination capacity in the selected water-scarce countries for the coming 40 years is summarized in Table 2.3. As illustrated in Table 2.3, the current total population in these 13 countries is 715 million which will be almost double (1,252 million) by 2050. Out of these, approximately 10 - 50 % population lives in urban cities. By 2050, a desalination capacity of 57.1 Mm^3/d is needed to maintain the current per capita water demand (0.146 $m^3/cap/d$) and to compensate the freshwater withdrawals. This indicates the growth of desalination market of 53.2 Mm^3/d, which is approximately 1,464 % increase as compared to the current installed capacity (3.9 Mm^3/d) in these 13 countries. However, there exist challenges for the implementation of the desalination technologies in these countries, which still need to be overcome.

Table 2.3: The current installed and projected desalination capacity, sea, and brackish water desalination plants m³/d, in the selected water-scarce countries

Country	N_{2015} [millions]	N_{2050} [millions]	Urban population [% share]	Q_{SW} [m³/d x1000]	Y_{SW}	Q_{BW} [m³/d x1000]	Y_{BW}	Q_{2016} [m³/cap/d x 10⁻³]	WW_{AVG} [m³/cap/d]	ΔQ_{2050} [Mm³/d]
Afghanistan	32.5	69.5	0.22	0.00	0	2.50	0.85	0.30	0.146	2.2
Burundi	11.2	19.5	0.11	0.00	0	0.00	0	0	0.146	0.3
Eritrea	5.2	10.5	0.21	1.00	1.00	0.15	0	0.92	0.146	0.3
Ethiopia	99.4	171	0.16	1.70	0.42	0.02	1.0	0.05	0.146	4.0
Malawi	17.2	25.9	0.19	0.00	0	0.00	0	0	0.146	0.7
Niger	19.9	53	0.17	0.00	0	0.00	0	0	0.146	1.3
Nigeria	182	258.5	0.49	10.90	0.55	4.70	0.15	0.08	0.146	18.5
Pakistan	189	348.7	0.36	44.60	0.35	93.10	0.02	0.26	0.146	18.3
Somalia	10.8	39.7	0.37	0.10	1.00	0.00	0	0.03	0.146	2.1
Sudan	40.2	60.1	0.39	43.60	0.81	0.48	0	2.25	0.146	3.4
Tanzania	53.5	69.1	0.26	0.60	1.0	6.10	0	0.04	0.146	2.6
Uganda	39.0	103.2	0.13	0.00	0	0	0	0	0.146	2.0
Zimbabwe	15.6	23.5	0.38	0.00	0	0	0	0	0.146	1.3
Total	715.5	1,252.2		102.5		107		3.9		57.1
Average	55.0	96.3		7.9		8.9		0.3		4.4

2.4   Case studies

A closer analysis of countries where a significant installed desalination capacity already exists such as India (2.2 Mm³/d), China (5.9 Mm³/d), Algeria (2.7 Mm³/d) and South Africa (0.24 Mm³/d) was performed (Figure 2.7). The analysis was carried out based on the use of raw water sources (seawater, brackish water or others), user categories (municipal, industrial or others), and the type of technology used, i.e., membrane (RO, ED, and NF), thermal (MSF, MED) or others.

Among the four selected countries, India and China currently have total available renewable water resources higher than the threshold level of water scarcity (1,000 m³/cap/year), which is distributed to a total population of approximately 1,350 million. However, the current trend of rural-urban migration and unequal distribution of the renewable water resources, regional water crisis is an issue in these two countries. This problem was the driving force for these countries to treat the unwanted water such as seawater, brackish water, and wastewater using desalination technology. The currently installed desalination capacity in India is 2.2 million m³/d, and in China is 5.9 million m³/d. In both countries, the membrane-based desalination leads the thermal-based desalination, and 75 % of the desalinated water is used for industrial purposes.

The other selected two countries such as South Africa and Algeria currently have total available renewable water resources below 1,000 m³/cap/year, distributed to a total population of approximately 50 million. As of today, Algeria suffers from a severe water scarcity, which forced them to be abstract all the available renewable water resources. However, South Africa on the other hand tried to maintain its available freshwater resources even though it has been categorized as a water scarce country (FAOAquastat., 2017). Both countries have started to treat the water using desalination technology to meet the growing water demand. As of 2016, the installed total desalination capacity is 2.7 million m³/d in Algeria and 0.24 million m³/d in South Africa. In Algeria, 90 % of the desalinated water is used for municipal purpose while in South Africa about 40 % of the desalinated water is used for municipal and 50 % for the industry.

Overall, the strategy adopted by India and China, i.e., to supply desalinated water to the industrial use could be applied to other countries to solve the issue of water scarcity by 2050.

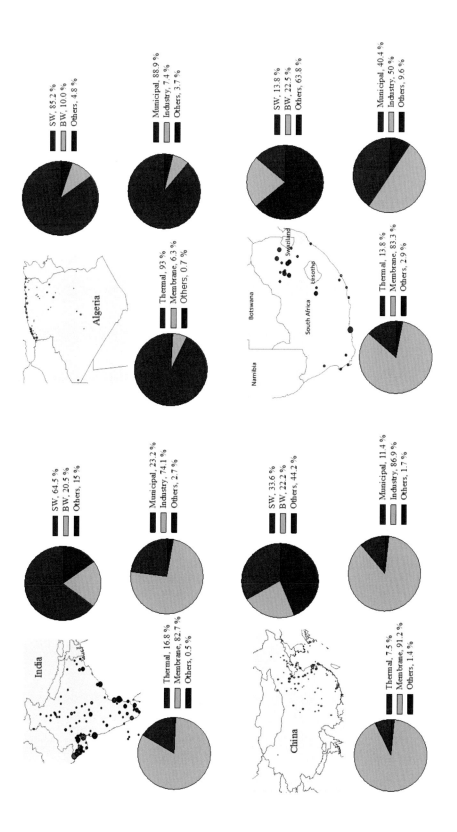

Figure 2.7: Country Comparison of desalination use in India, Algeria, China, and South Africa (DesalData, 2016)

2.5   What are the challenges?

Desalination is "*often chemically, energetically and operationally intensive, focused on large systems, and thus requires a considerable infusion of capital, engineering expertise, and infrastructure*" (Shanon et al., 2008). The main "Achilles heel" for the efficient operation of the membrane-based desalination systems is membrane fouling (Flemming et al., 1997). The problems associated with the membrane fouling are decreased membrane permeability, increased operating pressure, increased frequency of chemical cleaning and membrane deterioration (Matin et al., 2011). Despite all these facts, desalination is gaining a market to meet the demand of freshwater shortage worldwide. However, several factors that drive the feasibility of desalination plants need to be considered. Figure 2.8 indicates the drivers and restraints for desalination market. The significance of each factor is indicated by the length of the arrow as indicated in Figure 2.8. The dotted arrows highlight the forces whose importance is gradually decreasing (Nellen, 2011). As illustrated, the main drivers for desalination markets are a saltwater intrusion, the willingness of private investors to invest, water shortages and reduced plant prices, etc. While, on the other hand, the environmental impact, high capital cost, and political instability are main restraints to the desalination market.

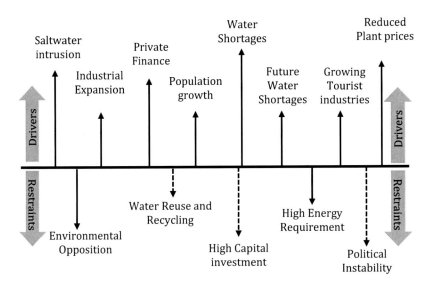

Figure 2.8: Key desalination market forces (Nellen, 2011)

2.5.1 Economic issue

Desalination is still the most energy-intensive technology to produce drinking water and is usually only implemented as a last resort where conventional freshwater resources stretched to the limit. The global concerns over climate change, water scarcity, rapid urbanization, and industrialization are some factor that led many scientists and engineers to think of desalination to meet the demand of freshwater supply worldwide (Gude, 2016). However, the cost needed to produce and distribute freshwater from seawater through desalination is still a matter of debate when compared with the cost associated with the conventional water supply systems (coagulation-flocculation-sedimentation and filtration scheme). As illustrated in Table 2.4, the cost of desalinated water was approximately 2 times higher compared to the cost of conventional water supply. Likewise, the cost related to energy consumption was also approximately 5 - 25 times higher for desalinated water compared to conventionally treated water (Voutchkov, 2011, 2014; Plappally, 2012) and cited by (Gude, 2016).

Table 2.4: Comparison of water costs for conventional & desalination water supply options (Voutchkov, 2011, 2014; Plappally, 2012) as cited by (Gude, 2016).

Range	Energy requirements (kWh/m^3)		Water production costs ($$/m^3$)	
	Conventional water supplies	Seawater reverse osmosis (SWRO)	Conventional water supplies	Seawater reverse osmosis (SWRO)
Low	0.1-0.5	2.5-2.8	0.25-0.75	0.5-0.8
Medium	1.0-2.5	3.0-3.5	0.75-2.50	1.0-1.5
High	2.5-4.5	4.0-4.5	2.50-5.00	2.0-4.0

In most of the urban cities, the available freshwater water resources have reached the capacity limit mainly linked to the population growth and urbanization. This circumstance has forced many cities to treat the brackish water, seawater, and wastewater via desalination or transport the freshwater from long distance. The choice depends upon the cost requirement or the willingness of the government of the respective countries. Gude (2016) compared the relative average cost for providing drinking water in various countries with conventional treatment (Figure 2.9) and transported desalination (Figure 2.10). For instances, people in Beijing, China pay the average cost of about $1.13/m^3$ for desalinated municipal water. The desalinated seawater in Beijing was collected from a distance of 135 km and has an elevation difference of 100 m from source to the distribution. Likewise, in Delhi, India the cost for the desalinated water is about \$ 1.9/m³. In this city, the desalinated water was transported from a distance of 1050 km and has an elevation difference of 500 m. On the other hand, the costs paid for municipal water by most of the European citizens are much higher compared to the cost paid by citizens from developing countries. The difference in the cost for water could be either due to government policy or due to strict environmental and economic standards of the country. For instance, the lower water prices in India and China compared to European countries could be related to the fact that in these countries the water prices are highly subsidized by the government (Gude, 2016).

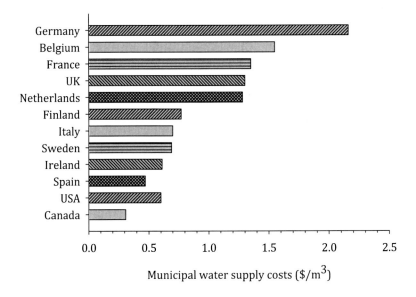

Figure 2.9: Comparison of municipal water costs treated with conventional treatment in different countries (Gude, 2016).

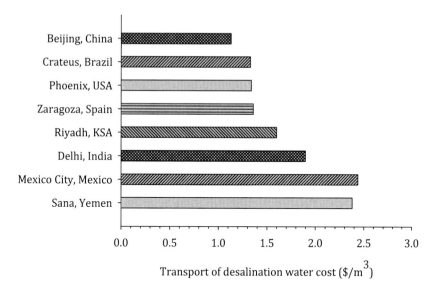

Figure 2.10: *Comparison of municipal water costs treated with transported desalination in different countries (Gude, 2016).*

2.5.2 **Environmental issue**

The desalination plants have a significant environmental impact. Despite many efforts, there are still some environmental concerns (Lattemann et al., 2008) such as;

- Disposal of material use,
- Land use
- Energy use to desalinate water and greenhouse gas (GHGs) emission,
- Discharge of concentrate
- High volume of chemical use
- Loss of aquatic organism from marine pollution and open seawater intake

The use of fossil fuels to desalinate the water emits the greenhouse gas, which includes carbon monoxide (CO), nitric oxide (NO, nitrogen dioxide (NO_2) and sulfur dioxide (SO_2). The recent technological advanced helped to decrease the emission of GHGs and depends upon if oil is used instead of natural gas (Dawoud et al., 2012). Likewise, the use of the high volume of chemicals during pre and post-treatment of seawater is another environmental concern. The main concern is the discharge of chemical into the natural water, which affects the ecological imbalance (Lattemann et al., 2008). Furthermore, the design of open seawater intake has a potential role in the loss of aquatic organism, as these organism colloids with the intake screen or sometimes drawn into the plant (Dawoud et al., 2012). Some of the possibilities for the sustainable solutions to prevent/minimize the issue listed above are (Lattemann et al., 2008);

- implement low or no chemical technologies
- treat the chemicals before discharging into the natural water bodies
- disperse the concentrate through a multiport diffuser in a suitable marine site
- use subsurface or submerged intakes with low intake velocities, and
- reuse of material

The summary of environmental challenges and possible sustainable solutions are illustrated in Figure 2.11.

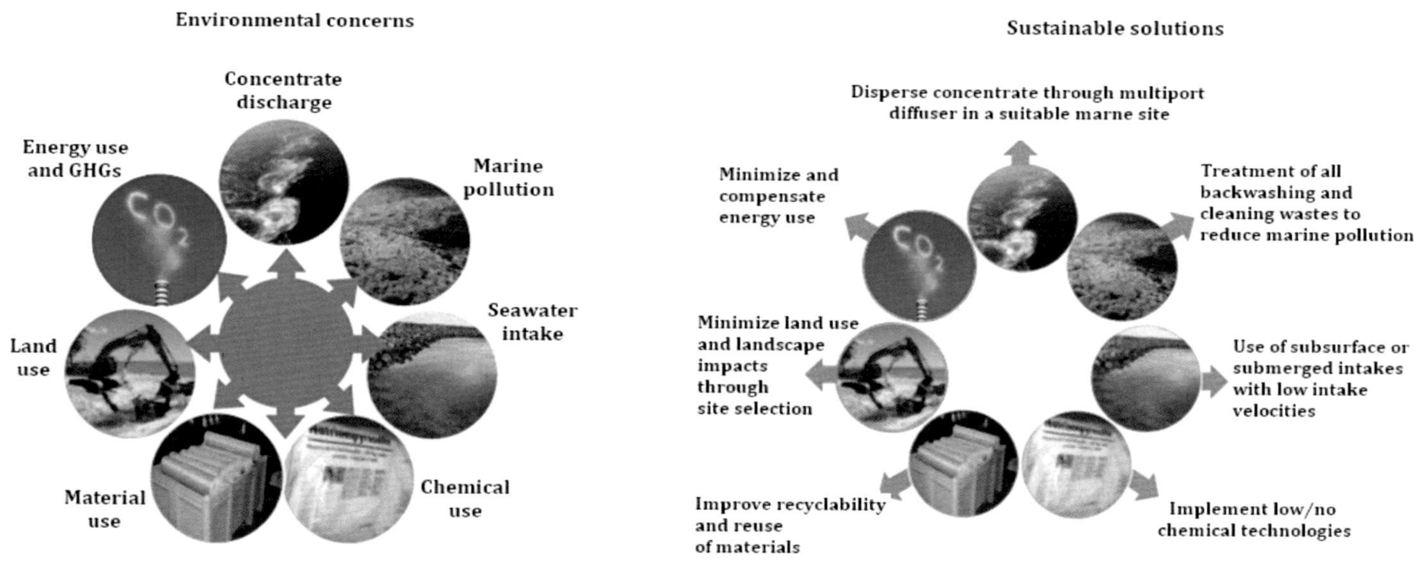

Environmental concerns

Marine pollution

Seawater intake

Concentrate discharge

Chemical use

Energy use and GHGs

Land use

Material use

Sustainable solutions

Disperse concentrate through multiport diffuser in a suitable marine site

Treatment of all backwashing and cleaning wastes to reduce marine pollution

Use of subsurface or submerged intakes with low intake velocities

Implement low/no chemical technologies

Minimize and compensate energy use

Minimize land use and landscape impacts through site selection

Improve recyclability and reuse of materials

Figure 2.11: Environmental concerns and sustainable solution for the desalination plant to minimize the environmental impact (Lattemann et al, 2008)

2.5.3 Membrane fouling

Membrane fouling is still the main "Achilles heel" for the cost-effective application of reverse osmosis (Flemming et al., 1997). The types of fouling are categorized into i) particulate/colloidal fouling, ii) inorganic fouling (Scaling), and iii) organic and biofouling. To prevent the occurrence of membrane fouling, SWRO plants are always equipped with pre-treatment systems (e.g., media filters with coagulation), MF/UF, etc. Moreover, the particulate and colloidal fouling are mostly controlled with this improvement in the pre-treatment; but the occurrence of organic and biofouling is still a major issue in SWRO membranes. The consequences of which are;

- Increase in head loss across the feed spacer of spiral wound elements
- Higher energy consumption to maintain the constant flux operation
- Higher chemical cleaning frequency
- Increase the replacement of membrane due to irreversible membrane fouling
- Decrease the rate of water production due to longer downtime during chemical cleaning and membrane replacement
- Increase salt passage and thus deteriorate the permeate quality

Reliable methods to monitor the membrane fouling potential of raw and pre-treated water is important in preventing and diagnosing fouling and to develop the effective fouling control strategies for the cost-effective operation of SWRO membranes. The most relevant and important parameters/indicators/methods are presented in Table 2.5. The details about indicators are described in the following chapters of this thesis.

Table 2.5: Relevant indicators/parameters to monitor the membrane fouling in SWRO membranes

Particulate fouling	Organic fouling	Biofouling	Others
- Silt density index (SDI) - Modified fouling index (MFI) - Modified fouling index (MFI-UF)	- Liquid chromatography organic carbon detection (LC-OCD) - Total organic/ dissolve organic carbon (TOC/DOC) - UV$_{254}$ - Fluorescence excitation and emission matrix (FEEM) - Fourier transform infrared spectroscopy (FTIR)	- Transparent exopolymer particles (TEP) - Assimilable organic carbon (AOC) - Adenosine triphosphate (ATP) - Biofouling potential (BFP) - Membrane fouling simulator (MFS)	- Algal cell concentration - Chlorophyll-a concentration

2.6 Concluding remarks

- By 2050, about 44 countries (2 billion people) will be strongly hit by water scarcity, and more than 95 %, i.e., 1.9 billion of these people may live in developing countries. The countries are Uganda, Burundi, Nigeria, Somalia, Malawi, Eritrea, Ethiopia, Haiti, Tanzania, Niger, Zimbabwe, Afghanistan, Sudan, and Pakistan.

- Currently, the majority of these 13 water-scarce countries have not yet installed desalination to meet their freshwater demand. The projected desalination capacity of 57 Mm³/d is needed by 2050 in these countries to meet the standard of current water demand and to compensate the withdrawal of renewable resources.

- The experience of some countries (e.g., India, China, and South Africa) in using desalination water to industrial users may be adopted in other nations to solve the issue of water scarcity by 2050.

- The current global trend showed that the desalination technology is finding new outlets as an alternative source for supplying water to meet growing water demand in most of the water-scarce countries. However, there have been barriers to its widespread adoption of technology mainly due to its cost, energy, lack of expertise, and the footprint.

2.7 References

Bremere, I., Kennedy, M., Stikker, A. and Schippers, J. (2001) How water scarcity will effect the growth in the desalination market in the coming 25 years. Desalination 138(1), 7-15.

Brown, A. and Matlock, M.D. (2011) A review of water scarcity indices and methodologies. The sustainability consortium Whitepaper 106.

Caron, D.A., Garneau, M.-È., Seubert, E., Howard, M.D.A., Darjany, L., Schnetzer, A., Cetinić, I., Filteau, G., Lauri, P., Jones, B. and Trussell, S. (2010) Harmful algae and their potential impacts on desalination operations off southern California. Water Research 44(2), 385-416.

Dawoud, M.A. and Al Mulla, M.M. (2012) Environmental Impacts of Seawater Desalination: Arabian Gulf Case Study. International journal of environmental and sustainability 1(3), 22-37.

DesalData (2016) accessed on: www. DesalData.com

Dhakal, N., Rodriguez, S.G.S., Schippers, J.C. and Kennedy, M.D. (2014) Perspectives and challenges for desalination in developing countries. IDA journal of desalination and water reuse 6(1).

Elimelech, M. (2012) Seawater Desalination, 2012 NWRI Clarke prize conference, Newport Beach, California.

FAOAquastat. (2017) http://www.fao.org/nr/water/aquastat/data/query/results.html accessed on 20 February 2017.

Flemming, H.C., Schaule, G., Griebe, T., Schmitt, J. and Tamachkiarowa, A. (1997) Biofouling—the Achilles heel of membrane processes. Desalination 113(2–3), 215-225.

Fragkou, M.C. and McEvoy, J. (2016) Trust matters: Why augmenting water supplies via desalination may not overcome perceptual water scarcity. Desalination 397, 1-8.

Gude, V.G. (2016) Desalination and sustainability – An appraisal and current perspective. Water Research 89, 87-106.

Lattemann, S. and Höpner, T. (2008) Environmental impact and impact assessment of seawater desalination. Desalination 220(1–3), 1-15.

Matin, A., Khan, Z., Zaidi, S.M.J. and Boyce, M.C. (2011) Biofouling in reverse osmosis membranes for seawater desalination: Phenomena and prevention. Desalination 281, 1-16.

Nellen, A. (2011) Desalination: A viable answer to deal with water crises? Future direction international.

Shanon, M.A., Bohn, P.W., Elimelech, M., Georgiadis, J.G., Marinas, B.J. and Mayes, A.M. (2008) Science and technology for water purification in the coming decades. Nature 452(20), 301-310.

Villacorte, L.O., Tabatabai, S.A.A., Dhakal, N., Amy, G., Schippers, J.C. and Kennedy, M.D. (2015) Algal blooms: an emerging threat to seawater reverse osmosis desalination. Desalination and Water Treatment 55(10), 2601-2611.

WaterReuseAssociation (2012) Seawater desalination costs, white paper.

Wilf, M. and Awerbuch, L. (2007) The Guidebook to Membrane Desalination Technology: Reverse Osmosis, Nanofiltration and Hybrid Systems, Process, Design, Applications, and Economics, Balaban Desalination Publications.

WorldBank (2017) http://databank.worldbank.org/data/reports.aspx?source=Health%20Nutrition%20and%20Population%20Statistics:%20Population%20estimates%20and%20projections#, accessed on 20 February 2017.

3

Measuring bacterial regrowth potential (BRP) in seawater reverse osmosis using a natural bacterial consortium and flow cytometry (FCM)

1. Contents

This chapter is based on:

Dhakal, N., Salinas Rodriguez, S.G., Ampah, J., Abushaban, A., Villacorte L.O., Schippers, J.C. and Kennedy, M.D. (2017), Measuring bacterial regrowth potential (BRP) in seawater reverse osmosis using a natural bacterial consortium and a flow cytometry. Submitted to Desalination.

Abstract

The study assessed bacterial regrowth potential (BRP) and flow cytometry (FCM) as a method for measuring the biofouling potential of seawater reverse osmosis (SWRO) feed water. The method involves the removal of bacteria from the water sample, followed by re-inoculation with a natural consortium of marine bacteria (10^4 cells/mL), with incubation (30 °C), and bacterial enumeration using FCM. Result illustrated that the BRP method is a potential tool to evaluate the biofouling potential of SWRO feed water. The method was relatively fast (2-3 days) compared to the conventional bioassays. The BRP method has the limit of detection of 43,000±12,000 cells/mL (9.3±2.6 μg-$C_{glucose}$/L), which was improved by introducing the heating of chemical (NaCl) and bottle at 550 °C for 6 hours during blank preparation. The method was calibrated with glucose as a standard substrate and showed a good linearity (R^2=0.88 to 0.95) between a range of 0-2,000 μg-$C_{glucose}$/L. The method was applied to monitor the reduction of bacterial regrowth potential by pre-treatment schemes of full-scale SWRO plants. Results showed that the DAF-UF as pre-treatment reduced the bacterial regrowth potential of SWRO feed water by 54 %, while it was 40 % with DMF-CF. Furthermore, the measured bacterial regrowth potential of SWRO feed water especially in the DAF-UF-RO scheme was found consistent with the frequency of chemical cleaning in SWRO systems. However, more investigation is still needed to demonstrate the link between the measured bacterial regrowth potential of SWRO feed water and the rate of biofouling development in SWRO systems.

Keywords: *Bacterial regrowth potential, net bacterial regrowth, flow cytometry, seawater reverse osmosis*

3.1 Introduction

Biofouling remains the major challenge for the cost-effective operation of seawater reverse osmosis (SWRO) membranes (Filloux et al., 2015, Flemming et al., 1997, Greenlee et al., 2009, Matin et al., 2011, Quek et al., 2015). Biofouling is triggered by the formation of a biofilm resulting from the growth of microorganisms on the membrane surface (Weinrich et al., 2016). In SWRO, the main consequences of biofouling are i) decreased membrane permeability, ii) increased pressure drop along the spacer channel, iii) increased the frequency of chemical cleaning and iv) possible increase in replacement frequency of membrane (Matin et al., 2011). In practice, several methods for biofouling control have been investigated such as; the application of the pre-treatment prior to SWRO to remove bacteria and biodegradable organic matter (Filloux et al., 2015), dosing of biocides (Kim et al., 2009), and limiting essential nutrients such as carbon, and phosphate (Jacobson et al., 2009, Vrouwenvelder et al., 2010). Detection of biofouling, while membranes remain in operation, is of great importance to avoid costly sacrifice of SWRO elements for autopsy (Dixon et al., 2012)

A very promising online detection of biofouling was attempted using membrane fouling simulator (MFS) (Vrouwenvelder et al., 2006); however, the rate of biofouling in MFS units occurs at the same rate as in the full-scale SWRO plant if fed with the same feed water. Thus, the detection of biofouling in an early stage using MFS is limited. Flow cytometry (FCM) and Bacterial Regrowth Potential (BRP) are two potential techniques that can provide information on the growth potential of the feed water to the SWRO while membranes remain in operation (Dixon et al., 2012). The conventional growth assays such as assimilable organic carbon (AOC) (Van der Kooij et al., 1982), and biodegradable dissolved organic carbon (BDOC) (Servais et al., 1987) also measures the bacterial growth; however it only estimate the concentration of growth-limiting compounds (e.g., AOC)(Prest et al., 2016). The previous studies have revealed that inorganic nutrients such as phosphorous to be the growth limiting compound in drinking water (Sathasivan et al., 1999), and in SWRO (Jacobson et al., 2009, Vrouwenvelder et al., 2010). Hence, if inorganic nutrients limit bacterial growth, the measurement with conventional thinking might lead to ineffective biofouling control measures (Sathasivan et al., 1999).

The BRP method has already been applied to assess the bacterial regrowth potential in water distribution systems. The BRP method aims to measure the potential of a water sample to support bacterial regrowth (Hambsch et al., 1993, Prest et al., 2016, 2013, Sathasivan et al., 1999, Withers et al., 1998). Recently, Dixon et al. (2012) published a paper that describes the application of BRP using flow cytometry and indigenous bacteria as inoculum for biofouling detection in SWRO membranes (Dixon et al., 2012). Moreover, the potential contribution of blank during BRP tests was not considered. Prest et al. (2016) reported bacterial regrowth up to 133 (±18) ×10^3 cell/mL in bottled water, a blank considered for BRP test in the freshwater sample (Prest et al., 2016). While, the blank for seawater sample should consider artificial seawater (ASW) prepared with same salinity to avoid the effect of an osmotic shock to marine bacteria (Csonka, 1989). Moreover, the contamination that might originate from the bottle, chemicals, pipette, and laboratory environment during ASW (blank) preparation could influence the result of FCM and BRP. Thus, it is essential to lower the interference of ASW, which also supports to achieve the lower limit of detection of the BRP method.

The BRP method in general consists of several steps, such as sample preparation, bacteria inoculation, and determination of the maximum bacterial regrowth potential. Sample preparation step for BRP includes the removal of indigenous bacteria. Removal of bacteria can be achieved by either filtration, pasteurization, and/or sterilization or combination of them as suggested by various growth assays (Hammes et al., 2005, Jeong et al., 2013, Kaplan et al., 1993, Le Chevalier et al., 1993, Sathasivan et al., 1999, Van der Kooij et al., 1982, Weinrich et al., 2016, Weinrich et al., 2011, Werner et al., 1986), but each method has advantages and disadvantages. For instance, the disadvantages of filtration method are i) release of organic carbon from filters, and ii) rejection of fraction of organic carbon. Likewise, the disadvantages of pasteurization and sterilization methods could be i) change of the organic carbon quality of sample (denaturation of proteins) (Ross et al., 2013), ii) precipitation of inorganic salts such as carbonate, phosphate, and iii) presence of dead or inactive bacterial cells in water sample. The dead or inactive bacterial cells may be a good source of carbon for the inoculated bacteria. It has been reported that one bacterial cell contains approximately 2 x 10^{-14} g of carbon (Batté et al., 2003).

The inoculum selected for BRP either a pure culture or a natural consortium of bacteria might affect the outcome of BRP tests. Rose et al. (2013) reported higher bacterial regrowth in fresh water using an indigenous bacterial consortium compared to a pure culture strain (Ross et al., 2013). The advantages of using a pure culture as inoculum are; i) the inoculum does not change over time and ii) results at different locations and time can be compared. The limitation of using a pure culture is that it can only utilize only certain biodegradable compounds (Servais et al., 1987) excluding polysaccharides. However, the use of a natural consortium broadens the substrate spectrum range (Ross et al., 2013, Withers et al., 1998). The standardization of bacterial yield factor while using natural consortium is difficult as it varies from location to location and even in time at the same location (Hammes et al., 2005)

Bacterial enumeration using FCM is a direct and fast method compared to conventional heterotrophic plate counting. FCM has been widely applied in freshwater(Hammes et al., 2005, Prest et al., 2013, Vital et al., 2012) and its application especially in seawater, has not been extensively studied. Recently, Dixon et al. (2012), and Van der Merwe et al. (2014) have explored the application of FCM in seawater samples (Dixon et al., 2012, Van der Merwe et al., 2014). Flow cytometry is capable of enumerating total bacterial cell counts (TCC) (Live+Dead) as well as distinguishing live and dead bacterial cells. Total bacterial cell counts can be enumerated using nucleic acid targeting stains such as DAPI or SYBR® Green I (Hammes et al., 2008, Zipper et al., 2004). While, the live and dead bacterial cells can be distinguished by staining with SYBR® Green I and Propidium Iodide (PI). The dye PI can only penetrate cells with disrupted membranes while SYBR® Green I can bind the nucleic acid of both live and dead bacterial cells (Stiefel et al., 2015). Two problems associated with FCM while enumerating marine bacteria are; i) distinction between the stained microbial cells and instrument noise, or sample background and ii) effect of salinity on staining dye. A research work performed by Prest et al. (2013) had proposed an electronic gate on the green/red fluorescence density plot to distinguish the signals of freshwater bacteria and instrument noise or sample background (Prest et al., 2013). However, the electronic gating for seawater samples needs to be verified or modified. Likewise, Barth et al. (2012) reported that salinity affects the ability of Propidium monoazide (PMA), an alternative dye to PI, to distinguish between live and dead cells in marine samples (Barth Jr et al., 2012), which needs to be tested for staining dye SYBR® Green I and Propidium Iodide (PI) used in this study.

The main aim of this study was to develop an improved method to measure the bacterial regrowth potential (BRP) using flow cytometry (FCM) and a natural bacterial consortium as inoculum. The BRP method studied was based on the method used by (Dixon et al., 2012) for seawater and (Prest et al., 2013) for freshwater. The BRP method was originally developed by (Sathasivan et al., 1999, Withers et al., 1998) for freshwater.

i)   To check the interference of salinity while enumerating seawater bacteria using FCM coupled with the fluorescence staining dye SYBR® Green I and Propidium Iodide (PI)

ii) To minimize the effect of blank (artificial seawater) and improve the limit of detection of the BRP method when applied to seawater samples

iii) To compare the effectiveness of filtration, pasteurization, and/or sterilization approach in removing or inactivating seawater bacteria from sample during BRP test

iv) To apply FCM and BRP techniques to monitor the bacterial regrowth potential of samples collected from the treatment process trains of full-scale SWRO plants.

3.2   Materials and methods

3.2.1 Artificial seawater (blank) preparation

Artificial seawater (ASW) was prepared based on the average inorganic ion concentration of coastal North Seawater (see supplementary data S1). To make up ASW, J.T. Baker analytical grade salts (Na_2CO_3, $NaHCO_3$, $CaCl_2.2H_2O$, KCl, Na_2SO_4, $MgCl_2.6H_2O$, NaCl) were sequentially dissolved in MilliQ water. The MilliQ water was produced from tap water purified via a series of treatment steps: reverse osmosis > electro deionization > granular activated carbon adsorption > Ultraviolet (UV) disinfection > 0.22 μm filtration.

3.2.2 Glassware preparation

All the tests were performed in glassware/vials, which were cleaned according to the protocol described by (Prest et al., 2013). In short, the glassware and caps were first washed with detergent and then rinsed three times with MilliQ water. They were then soaked overnight in 0.2 M HCl solution, rinsed three more times with MilliQ water, and air-dried. Finally, all the glassware was heated in a muffle furnace at 550 °C for 6 hours, to remove all traces of organic material. The caps were soaked in 100 g/L sodium persulfate solution ($Na_2S_2O_8$) at 60 °C for 1 hour, rinsed three times with MilliQ water and air-dried. The glassware was tightly covered with the cap.

3.2.3 Bacterial growth medium preparation

All nutrient stock solutions of organic carbon (sodium acetate, 100 mg/L, glucose, 100 mg/L), and inorganic nutrients such as phosphorous as ($NaH_2PO_4.2H_2O$, 200 mg/L) and nitrogen as ($NaNO_3$, 25 mg/L) were prepared using Milli-Q water in separate clean glass bottles. All bottles were capped tightly, covered with aluminum foil, mixed in a shaker at 150 rpm for 1 hour, and then pasteurized (70 °C, 30 minutes). The nutrient stock solution was kept in a refrigerator at 4 °C and used during the calibration of BRP.

3.2.4 Bacterial inoculum

A natural bacterial consortium collected from the North Sea (Jacobahaven, the Netherlands) was used as a bacterial inoculum. The collected seawater sample was first pre-filtered onsite through a 2 μm glass filter to remove suspended particles and algae that were present in the seawater.

3.2.5 Analytical techniques/methods

3.2.5.1 Flow cytometry (FCM)

Bacterial enumeration was performed using an Accuri™ C6 flow cytometer (BD Biosciences, San Jose, CA). It was equipped with a blue laser that emits at a wavelength, $\lambda = 488$ nm, and four optical detectors for detecting green fluorescence (FL1 at $\lambda_{max} = 533$ nm), red fluorescence (FL3 at $\lambda_{max} = 670$ nm) and forward or side scatter (FSC or SSC). The calibration of the FCM during the enumeration was performed using partec calibration beads (3 μm) and 8 and 6 peaks calibration beads (see supplementary data S4).

Sample staining and bacterial enumeration

A protocol developed by (Prest et al., 2013) was used for sample staining and bacterial enumeration. In short, the sample (V = 500 µL) was preheated for 5 minutes at 35 °C and stained with SYBR Green®I+Propidium Iodide (PI). The stained sample was then incubated in the dark for 10 minutes at 35 °C before enumeration of the bacterial cells by FCM. Samples containing bacterial cells ≥10^7 cells/mL were diluted with 0.22 µm filtered ASW before staining as the bacterial cell concentrations should range from 10^2 to 10^7 cells/mL for FCM (Gatza et al., 2013). The same protocol was followed for all sample measurements.

Electronic gating

Electronic gating as shown in Figure 3.1 was used for the seawater sample. To establish this gate, seawater samples (collected from Jacobahaven, the Netherlands), were autoclaved at 121 °C for 20 minutes. The autoclaved samples were stained with SYBR Green®I+Propidium Iodide (PI), and bacterial enumeration was performed using FCM. After FCM counting, the blue gating (as shown in Figure 3.1) was adjusted until no live bacterial cells were found inside the gate and all dead cells, sample background and instrument noise fell outside the gate. The electronic gate was used throughout this study. A comparison result with various electronic gates is presented in supplementary data S5.

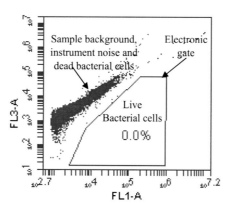

Figure 3.1: Gating strategy for marine bacterial cell selection using a fixed electronic gate on green fluorescence (FL1-A) and red fluorescence (FL3-A). The sample was autoclaved seawater stained with SYBR Green I and Propidium Iodine (PI).

3.2.5.2 Bacterial regrowth potential (BRP)

A BRP method in seawater sample was based on the method developed for freshwater by (Prest et al., 2013, Sathasivan et al., 1999, Withers et al., 1998). The BRP method comprises three steps: sample preparation, inoculation, and enumeration of bacterial regrowth. A scheme of the method is shown in Figure 3.2.

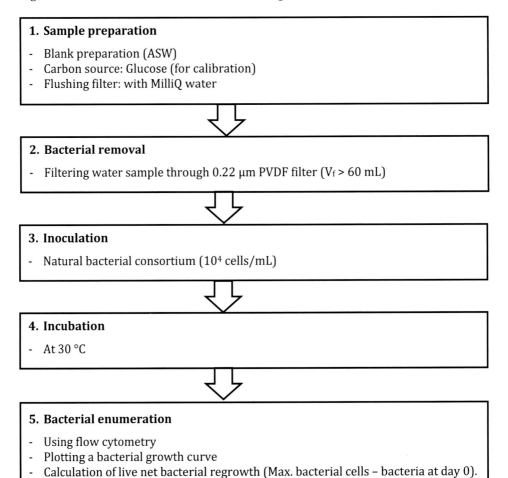

1. Sample preparation
- Blank preparation (ASW)
- Carbon source: Glucose (for calibration)
- Flushing filter: with MilliQ water

2. Bacterial removal
- Filtering water sample through 0.22 µm PVDF filter (V_f > 60 mL)

3. Inoculation
- Natural bacterial consortium (10^4 cells/mL)

4. Incubation
- At 30 °C

5. Bacterial enumeration
- Using flow cytometry
- Plotting a bacterial growth curve
- Calculation of live net bacterial regrowth (Max. bacterial cells – bacteria at day 0).

Figure 3.2: Procedure for BRP method using flow cytometry

The protocol involves filtration of a sample (V > 60 mL) through 0.22 µm PVDF filters to remove large particles and bacteria from seawater samples. Before the sample filtration, 0.22 µm filter was flushed with MilliQ water to remove the released carbon from the filter.

The choice of 0.22 μm filtration approach for BRP test was based on the comparative study made with different approaches as described in section 3.2.5.2.3. The 0.22 μm filtered seawater sample (20 mL) was transferred into clean vials (in triplicate). A volume of 200 μL (equivalent to 10^4 live cells/mL) of collected seawater from the North Sea (Jacobahaven, the Netherlands) was added to the vial containing 20 mL of sample (Withers et al., 1998). Samples were then incubated at a temperature of 30 °C (Prest et al., 2013) to achieve rapid bacterial regrowth and to reduce the time required to reach the maximum regrowth. A sample volume of 500 μL was taken from the incubated vials every 24 hours, and the live bacterial cell concentration was enumerated using flow cytometry according to the protocol described in section 3.2.5.1. The bacterial growth curves were then plotted, and the net live bacterial regrowth was calculated by subtracting the bacterial cell numbers at day zero (N_0) from the maximum bacterial count (N_{max}) during the incubation period. The net bacterial regrowth was considered as an indicator for BRP.

3.2.5.2.1 Effect of blank (ASW) and limit of detection of BRP method

The effect of blank (ASW) on BRP method was investigated. A protocol was proposed to minimize the effect of blank and improved the limit of detection of BRP method. It was hypothesized that possible contamination that might originate from bottles, chemicals used during the preparation of blank could be minimized by burning chemicals and bottles in a muffle furnace at 550 °C for 6 hours. To demonstrate this, different experimental conditions, as shown in Table 3.1, were tested and compared. In all cases, heating of NaCl was only considered due to its higher melting point (> 550 °C) and being the primary salts during blank preparation. In each test, blank sample was inoculated with a natural bacterial consortium, and its regrowth was monitored as explained in section 3.2.5.2.

Table 3.1: Preparation of blank (ASW) samples with different conditions

ASW I (All salts *) pH = 7.8, TDS = 35 g/L	ASW II (with only NaCl) pH = 5.5, TDS = 35 g/L	ASW III (with NaCl+NaHCO₃) pH = 7.5, TDS = 35 g/L
No heating of bottle and chemicals	Heating of both bottle and chemical (550 °C, 6 h)	Heating of both bottle and chemical (only NaCl) at 550 °C, 6 h.
Heating of bottle (550 °C, 6 h) and no heating of chemical		
Heating of both bottle and chemical (only NaCl) at 550 °C, 6 h		

All salts refers to NaCl, MgCl₂.6H₂0, Na₂SO₄, CaCl₂.2H₂0, KCl, NaHCO₃, Na₂CO₃

3.2.5.2.2 Effect of salinity on bacterial enumeration by FCM

To demonstrate the effect of salinity while enumerating seawater bacteria using FCM, the ASW (TDS = 35 g/L) was prepared. The ASW was then pre-filtered with 0.22 µm PVDF filter to remove any bacterial cells presents in ASW. The filtered ASW was then diluted with MilliQ water to have samples with a different salt concentration that ranged from 2 to 35 g/L. A sample volume of 30 mL was transferred into a clean vial (in triplicate) for all samples and inoculated with the same concentration of a natural bacterial consortium. The samples were then vortexed for 5 seconds, and live bacterial cells were enumerated and compared.

3.2.5.2.3 Effect of sample filtration, pasteurization, sterilization

For this purpose, seawater was collected from Jacobahaven, the Netherlands. The collected seawater samples were i) filtered with 0.22 µm, ii) filtered with 0.1 µm, iii) pasteurized (70 °C, 30 minutes), and iv) sterilized (121 °C, 20 minutes). All samples after each pre-treatment steps were inoculated with 10^4 cells/mL of a natural bacterial consortium, incubated at 30 °C and enumerated the bacterial regrowth over time using FCM. In all cases, the net live bacterial regrowth was calculated and compared.

3.2.5.2.4 Calibration of BRP methods

For calibration curve, glucose was added to a different range of concentration (0 – 2,000 µg-$C_{glucose}$/L) in both ASW and natural seawater. In addition to glucose, all the samples prepared with ASW were also spiked with a fixed concentration of 500 µg /L of N (NaNO₃) and 100 µg/L of P (NaH₂PO₄). The calibration curves were plotted between the live net bacterial regrowth over the concentration range of glucose (0 – 2,000 µg-C$_{glucose}$/L). The yield factor of the test bacteria (an inoculated natural bacterial consortium) was calculated from the slope of the calibration curves produced with ASW and natural seawater. The equivalent carbon concentration was then calculated using the Equation 3.1 as described by (Hammes et al., 2010).

$$\text{ECC } [(\mu\text{g C})\text{L}^{-1}] = \frac{\text{Net bacterial growth (cells L}^{-1})}{\text{Bacterial specific yield (Cells } \mu\text{g}^{-1})} \qquad \text{Equation 3.1}$$

3.2.6 Application of BRP method

The BRP method was applied to monitor the bacterial regrowth potential along the treatment process trains of full-scale desalination plants located in the Middle East. The general scheme of the plants included: i) dissolved air flotation/ultrafiltration/reverse osmosis (DAF-UF-RO) and ii) dual media filtration/cartridge filter/reverse osmosis (DMF-CF-RO) as shown in Figure 3.3 a, and b, respectively. Samples (S1, S2, S3, S4, S5, S6) as indicated by red dots in Figure 3.3a and b were simultaneously taken in a clean glass bottle and measured the bacterial regrowth potential using the BRP method.

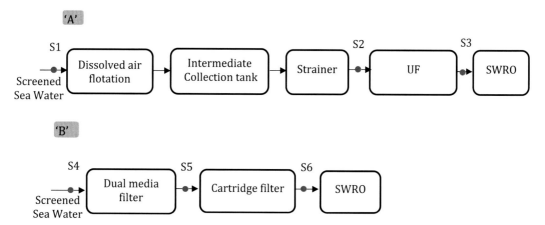

Figure 3.3: General scheme of a) the DAF-UF-RO plant and b) the DMF-CF-RO plant (red dots are sampling points)

3.3   Results and discussion

3.3.1 The use of FCM on enumerating seawater bacteria during BRP

The application of FCM on enumerating seawater bacteria during BRP tests was evaluated in this study. Figure 3.4a shows the reproducibility of the FCM used in this study, as determined by the serial dilution of the natural seawater bacteria in 0.22 μm PVDF filtered artificial seawater. A good linear relationship (R^2 = 0.99) between the percentage of seawater and live bacterial cell concentrations was observed, with a percentage deviation that ranged from 0.6-9.1 % (see Supplementary data S2). Testing was also performed to check the interference of salinity while enumerating seawater bacteria using FCM combined with fluorescence staining dye SYBR® Green I (SG) and Propidium Iodide (PI). Previous studies reported that salinity might affect the efficiency of staining dyes when distinguishing viable cells. More importantly, Barth et al. (2012) found out that the salinity affects the ability of Propidium monoazide (PMA), an alternative dye to PI, in distinguishing viable cells in marine samples (Barth Jr et al., 2012). It was described that the high concentrations (> 5 %) of sodium chloride (NaCl) prevented PMA from inhibiting DNA amplification from dead cells (Barth Jr et al., 2012). Moreover, this study showed no interference of salinity during the enumeration of seawater bacteria using FCM combined with fluorescence staining dye SYBR® Green I (SG) and Propidium Iodide (PI). As illustrated in Figure 3.4b, there was no substantial difference observed in the measured live bacterial cells concentration when same concentration of seawater bacteria was inoculated in AWS (TDS ranged from 15 - 35 g/L). The live bacterial cell concentration declined at a rate of 2,420 cell per g/L for TDS < 15 g/L, which could be attributed mainly to the effect of osmotic shock, which occurs when there is a sudden change in the solute concentration around bacterial cells. At a low level of salt concentration, water enters through the bacterial cells causing it to swell and finally burst (Csonka, 1989). The result elucidates the importance of diluting the seawater samples with same salinity artificial seawater (ASW) to avoid the effect of osmotic shock wherever necessary during FCM enumeration. Moreover, the organic carbon contamination that might originate from the bottle, chemicals, pipette, and laboratory environment during ASW (blank) preparation could influence the result of

FCM and BRP. Thus, it is rather important to lower the effect of ASW, which also supports to achieve the lower limit of detection of the BRP method.

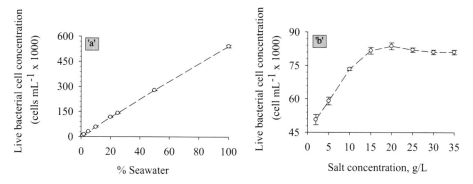

Figure 3.4: (a) Reproducibility and precision of flow cytometry for measuring live bacterial cell concentration in seawater sample and (b) Bacterial live cell enumeration by FCM combined with fluorescence staining dye SYBR® Green I (SG) and Propidium Iodide (PI) for a ASW sample prepared at various concentrations (0 - 35 g/L) and inoculated with the same concentration of seawater bacteria

3.3.2 Measurement of BRP in seawater sample

3.3.2.1 Effect of blank (ASW) and limit of detection of BRP method

The effect of heating of bottle and chemical at 550 °C for 6 hours was tested with the aim to lower the effect of blank and achieve the lower limit of detection of the BRP method. To demonstrate this, blank prepared with five different conditions as described in section 3.2.5.2.1 were tested. In all cases, the live net bacterial regrowth of the natural bacterial consortium was calculated and compared as shown in Figure 3.5. As illustrated in Figure 3.5 the consortium of bacteria proliferated much more in sample A (ASW prepared with all non-heated salts in a non-heated bottle) with the record of live net bacterial regrowth approximately 600,000 ± 65,000 cells/mL. The higher regrowth could be attributed to the organic contamination from chemicals and bottles used during the preparation of ASW. To prove this, another test was performed with ASW prepared with all non-heated salts in a heated bottle at 550 °C for 6 hours (Sample B). The result revealed that the live net bacterial regrowth was approximately 16 % lower compared to that measured in sample A. Furthermore, to see the effect of the contaminant from chemicals, another test was performed with ASW prepared with all salts where NaCl and the bottle were heated at 550 °C for 6 hours (Sample C). The measured live net bacterial regrowth in sample C was approximately 90 % lower compared to that measured in sample A. The result illustrates that heating of the major salt (NaCl) and bottle in a muffle furnace at 550 °C for 6 hours during ASW preparation substantially reduced the bacterial regrowth.

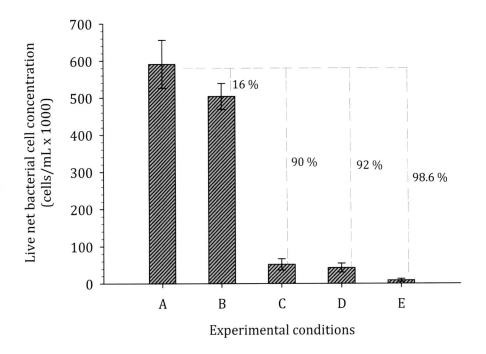

Figure 3.5: Comparison of live net bacterial regrowth in various ASW samples prepared with A) All salts (pH = 7.8), in which both bottle and salts were not heated. B) All salts (pH = 7.8) in which bottle was heated and salts were not heated. C) All salts (pH = 7.8) in which both bottle and salt (NaCl only) were heated. D) Two salts (NaCl + NaHCO₃) in which NaCl and bottle were heated, E) One salt NaCl, in which both NaCl and bottle were heated

Furthermore, the study observed the lowest live net bacterial regrowth (8,400 ± 4,000 cells/mL) in an ASW sample prepared with only NaCl (no other chemicals added), where both chemical (NaCl) and bottle were heated at 550 °C for 6 hours (sample E). This level was about 98.6 % lower than measured in sample A. Moreover, the measured lowest bacterial regrowth could be due to the low pH (5.5) of the sample and having no buffer capacity. Therefore, another ASW were prepared using two inorganic salts NaCl and NaHCO$_3$ where NaCl and bottle were both heated at 550 °C for 6 hours (Sample D). The result revealed the live net bacterial regrowth approximately to the level of 43,000 ± 12,000 cells/mL. This level was about 92 % lower than measured in sample A, which also indicates the lowest detection limit of the BRP method. Previous studies suggested that potassium, magnesium, and calcium are also essential elements required for the growth of marine bacteria (Tsueng et al., 2010, Unemoto et al., 1973). Moreover, we observed no significant difference in net bacterial regrowth with or without the addition of magnesium and calcium when compared net bacterial regrowth in sample C and D (Figure 3.5). Therefore, to avoid possible chemical contamination, ASW was prepared with only two salts: NaCl and NaHCO$_3$, where chemical (NaCl) and bottles are heated at 550 °C for 6 hours.

3.3.2.2 *Effect of sample filtration, pasteurization, sterilization*

The effectiveness of filtration, pasteurization (70 °C, 30 minutes), and sterilization (121 °C, 20 minutes) for removing or inactivating seawater bacteria were investigated. Results showed that the measured bacterial regrowth after pasteurization (70 °C, 30 minutes) was approximately 15 %, 27 %, and 34 % higher compared to samples after 0.22 µm filtration, sterilization at 121 °C, 20 minutes, and 0.1 µm filtration, respectively (Figure 3.6a). However, all the mentioned approaches have advantages and disadvantages. For instance, the disadvantages of pasteurization and sterilization could be i) change the organic carbon quality of sample (denaturation of proteins) (Ross et al., 2013), ii) precipitation of inorganic salts such as carbonate, phosphate, and iii) presence of dead or inactive bacterial cells in a water sample. The dead or inactive bacterial cells may be a good source of carbon for the inoculated bacteria. It has been reported that one bacterial cell contains approximately 2×10^{-14} g of carbon (Batté et al., 2003). Likewise, the disadvantages of filtration approach could be i) releases of carbon from the filter, and ii) rejection of fraction of organic carbon. Despite the variation in the live net bacterial regrowth, the BRP method studied used 0.22 µm filtration approach to remove the bacteria from water sample. The filtration approach was also recommended by (Weinrich et al., 2011, Withers et al., 1998) for bacterial removal from water sample. Although during the filtration, the virgin filter release the carbon but the flushing of the filters with MilliQ water substantially reduced the release of carbon (Figure 3.6b).

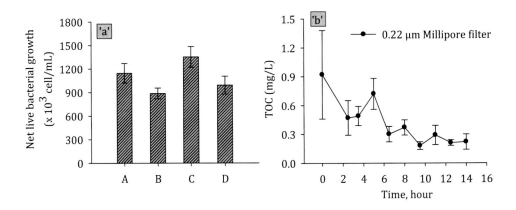

Figure 3.6: a) Comparison of the net live bacterial regrowth measured after filtration (A = 0.22 µm, B = 0.1 µm filter, C = pasteurization (70 °C, 30 minutes), and D = sterilization (121 °C, 20 minutes) and b) TOC released by 0.22 µm when flushed with MilliQ water

3.3.2.3 Calibration of BRP with glucose in ASW and natural seawater

The calibration of the BRP was performed in ASW (Figure 3.7), and natural seawater (Figure 3.8) fortified with 0 – 2,000 µg-$C_{glucose}$/L. Before the selection of glucose as substrate, a comparative study of bacterial regrowth using glucose and acetate as a substrate was performed. The result revealed that the response of a natural consortium of seawater bacteria to the utilization of acetate was approximately 3-fold higher than to the utilization of glucose (see supplementary data S3) with the linear fit, i.e., R^2 = 0.99 (Glucose) and R^2 = 0.96 (acetate). Despite this result, glucose was chosen as the standard substrate for the calibration of BRP. It has been reported that the glucose is a useful substrate during the characterization of a complex natural seawater microbial population (Vaccaro et al., 1968).

The calibration of BRP produced using glucose as a substrate in ASW, and natural seawater are presented in Figure 3.7 and 3.9, respectively. In both cases, the live bacterial net regrowth was determined from the growth curve and plotted against glucose concentration. As illustrated in Figure 3.7 and 3.9, the response of an inoculated natural bacterial consortium to the utilization of glucose as a carbon source showed a linear fit with R^2 > 0.95. The specific yield of inoculated bacteria, which is determined from the slope of the calibration curve, was approximately (4.4 - 4.6) ×10⁶ cells/μg-C. This was within the reported theoretical bacterial yield for P-17 (4.1 ×10⁶ CFU /μg acetate-C) and NOX (1.2 ×10⁷ CFU/μg acetate-C) (Hammes et al., 2005). Using the calculated specific bacterial yield and the net bacterial regrowth, the equivalent carbon concentration (ECC) was calculated according to the Equation 3.1 in both samples prepared with ASW and natural seawater (Table 3.2). Accordingly, the lowest measured value (LMV) for BRP method was 43,000 ± 12, 000 cells/mL, which is equivalent to a carbon concentration of 9.3 ± 2.6 μg-C$_{glucose}$/L. The measured LMV for BRP method is slightly lower than 10 μg-C acetate/L, a threshold value beyond which the biofouling is expected in freshwater (Van der Kooij et al., 1982). The measured ECC in natural seawater when C = 0 was approximately 817 μg-C$_{glucose}$/L. The higher measured ECC could be attributed to the occurrence of algal blooms in the seawater where the sample was collected.

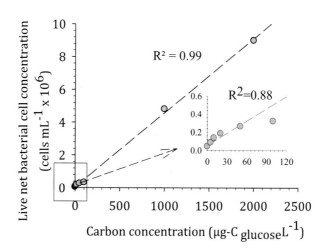

Figure 3.7: The correlation between the live bacterial net regrowth measured and carbon concentration (0 - 2,000 μg-C$_{glucose}$/L) added to the ASW sample.

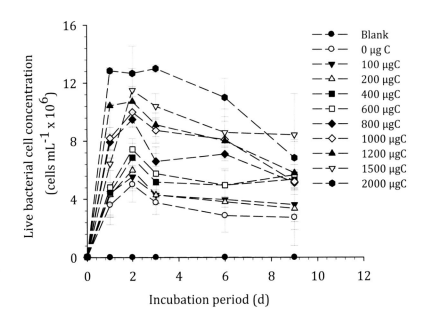

Figure 3.8: Monitoring the growth of the live bacterial cell in natural seawater samples when spiked with various carbon concentrations (0 - 2,000 µg-C$_{glucose}$/L)

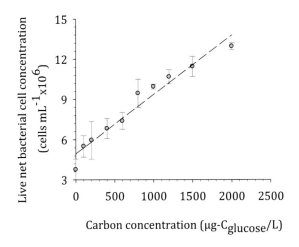

Figure 3.9: Correlation between the live bacterial net regrowth measured and various carbon concentration (0 - 2,000 µg-C$_{glucose}$/L) added to the natural seawater sample

Table 3.2: Maximum bacterial cell concentration (N_{max}) and ECC in artificial and natural seawater

Carbon concentration, $\mu g\text{-}C_{glucose}/L$	Artificial seawater		Natural seawater	
	Maximum live net bacterial cell, N_{max} ($\times 10^4$) (cells/mL)[a]	Equivalent carbon conc. ($\mu g C_{glucose}/L$)	Maximum live net bacterial cell, N_{max} ($\times 10^5$) (cells/mL)[a]	Equivalent* carbon conc. ($\mu g C_{glucose}/L$)
0	4.3 ± 0.1	9.3	37.6 ± 8	817
5	8.6 ± 0.07	18.7	n.m.	n.m.
10	13.8 ± 1.13	30.0	n.m.	n.m.
20	18.7 ± 1.84	40.6	n.m.	n.m.
50	26.5 ± 1.06	57.6	n.m.	n.m.
100	32.2 ± 4.60	70.0	55.5 ± 8	443
200	n.m.	n.m.	60 ± 14	546
400	n.m.	n.m.	68.0 ± 7.4	728
600	n.m.	n.m.	74.3 ± 6.2	871
800	n.m.	n.m.	95 ± 10.6	1,341
1,000	480.6 ± 1.84	1,045	100 ± 1.8	1,455
1,200	n.m.		107 ± 5.2	1,614
1,500	n.m.		115 ± 7.5	1,796
2,000	899.6 ± 7.00	1,956	130 ± 2.5	2,137

[a] values are average ± standard deviation; n = 3, n.m = not measured

* The calculated ECC for natural seawater is blank corrected.

3.3.2.4 Monitoring BRP in water treatment process trains of SWRO plants

Bacterial regrowth potential was monitored over the treatment process trains of two desalination plants, which included DAF-UF-RO, and DMF-CF-RO. The general schemes of the plants are shown in Figure 3.3 in section 3.2.6. Both plants abstract the raw water through an open intake of about 7 meters below the seawater surface, but from two different locations. In both intakes, shock chlorination (approximately 1 mg/L) was applied three times a day. The measured raw water pH was~8.55 which was adjusted to approximately 7.90 (DAF-UF-RO) and 7.4 (DMF-CF-RO) by dosing H_2SO_4 in both plants. The coagulant ($FeCl_3$) was continuously dosed in both plants at a concentration of 0.5 ppm of $FeCl_3$ before DAF and 0.8 ppm of $FeCl_3$ before DMF. The de-chlorination was performed before the SWRO unit by dosing $Na_2S_2O_5$ (sodium-meta-bisulfite). The dosing pump for $Na_2S_2O_5$ was controlled based on Oxidation-Reduction Potential (ORP) value, which was set to a level of 250 mV.

The monitored bacterial regrowth potential of samples collected from two process schemes are presented in Figure 3.10. From the growth curve, the net live bacterial regrowth supported by each sample was calculated to determine the percentage reduction towards bacterial regrowth potential by each treatment steps (Table 3.3). The measured bacterial regrowth potential of raw water of DAF-UF-RO treatment scheme, i.e., sample before DAF was approximately 1.6 folds higher than raw water of DMF-CF-RO line, i.e., sample before DMF. This variation could be due to the difference in raw water quality, as the abstractions of raw water were from two different intake locations. The result of Liquid chromatography organic carbon detection (LC-OCD) of samples performed showed that raw water of DAF-UF-RO has slightly higher biopolymer fraction (>20 kDa), i.e., 0.17 mg-C/L compared to the raw water of DAF-CF-RO, i.e., 0.12 mg-C/L. The difference in biopolymer fraction could be one of the reasons for higher bacterial regrowth. Passow (2002) also reported that the biopolymer fraction could be degraded by bacteria in a matter of few hours to several months (Passow, 2002).

The BRP result suggested that the DAF-UF performs better compared to DMF-CF, as DAF-UF reduced the bacterial regrowth potential of SWRO feed water by 54 %, while it was only 40 % by DMF-CF. This could be due to the higher removal of the biodegradable organic matter in dissolved air flotation operated with 0.5 mg/L of $FeCl_3$ coagulant and followed by ultrafiltration. Moreover, in terms of an absolute number of bacterial regrowth, the SWRO feed water after DAF and UF supports for 1.5×10^6 cells/mL, which is 1.25 times higher than in SWRO feed water of DMF-CF as illustrated in Figure 3.10b,d. This suggested that SWRO after DAF-UF is more vulnerable to biofouling compared to SWRO after DMF-CF. However, it should be noted that the raw water for DMF-CF had a lower net bacterial cell number to start with.

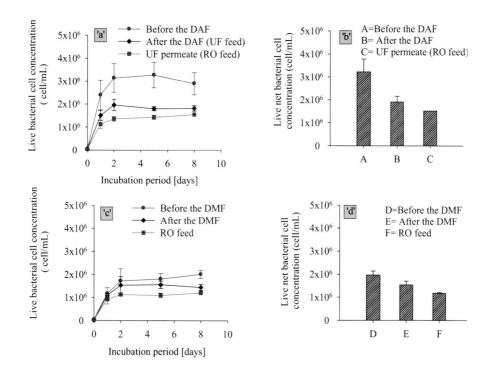

Figure 3.10: Live bacterial regrowth curve and calculated live net bacterial cell concentration from the growth curve for samples collected from a,b) DAF-UF-RO, and c,d) DMF-CF-RO treatment trains

Table 3.3: Live net bacterial growth measured using Flow Cytometry

Water samples	Live net bacterial regrowth (10⁶ cells/mL)	BRP reduction (%)
DAF-UF-RO scheme		
Before the DAF	3.3	
After the DAF	1.9	42
UF permeate (RO feed)	1.5	54
DMF-CF-RO scheme		
Before the DMF	2.0	
After the DMF	1.6	20
After cartridge filter (RO feed)	1.2	40

Nevertheless, other water quality parameters measured for SWRO feed water of DAF-UF-RO and DMF-CF-RO scheme during the time of the study showed a very low fouling potential as shown in Table 3.4. However, a remarkable increase in pressure drop was observed on the SWRO of the DAF-UF-RO systems, which demands higher frequency of cleaning-in-place (CIP). While, the SWRO in this scheme DMF-CF-RO was operated with no CIP for more than a year. Therefore, the historical SWRO operational data of DAF-UF-RO were collected and analyzed to answer the mystery of higher increase in pressure drop (Figure 3.11).

Table 3.4: RO feed water quality measured regarding SDI, MFI-UF, LC-OCD, and TEP

Parameters	units	RO feed water of	
		DAF-UF-RO	DMF-CF-RO
Silt Density Index (SDI)		< 1,7	~1,5
Membrane fouling potential (MFI-UF$_{10kDa}$)	s/L^2	690	bdl
Biopolymer concentration	mg-C/L	0,09	0,1
Transparent exopolymer particles (TEP$_{10kDa}$)	mgXeq/L	0,06	0,04

As illustrated in Figure 3.11, the increase in pressure drop (ΔP) in SWRO units shows a steady growth after each CIP performed, and currently is at a level that only a marginal increase in ΔP requires CIP again. It is a clear indication of biofouling in RO membranes because particulate fouling can be excluded due to the pre-treatment of the feed water with ultrafiltration as demonstrated by the measured SDI and MFI-UF values of RO feed water (Table 3.4). The remaining high head loss after CIP is the main reason for the relatively rapid increase in head loss and arriving rapidly at a level, which requires CIP to be repeated. The fundamental reason for the occurrence of biofouling in the membranes is the presence of biodegradable organic matter (Weinrich et al., 2013). The biodegradable organic matter could have been introduced into the treatment process by the impure chemicals dosed, namely sulfuric acid, sodium-meta-bisulfite, ferric (coagulant). These compounds might have been biodegraded in the DMF of the DMF-CF-RO scheme, but not in the DAF-UF-RO scheme. The DAF-UF-RO scheme has no treatment step, which incorporates biodegradation on non-polymeric compounds. There is evidence that biologically active sand filtration RO significantly enhanced the RO membrane performance (Griebe et al., 1998), presumably by removing the biodegradable organic matter (Weinrich et al., 2013). Also, it is possible that biodegradable organic compounds soon after the start-up of the DAF-UF-RO plant were present in the raw water, and were not adequately removed in the DAF-UF-RO scheme due to the absence of a treatment process incorporating biodegradation. These biodegradable compounds probably originated from algal bloom(s). The previous study also demonstrated a higher bacterial growth potential in a sample which has the higher concentration of algal organic matter (Villacorte, 2014).

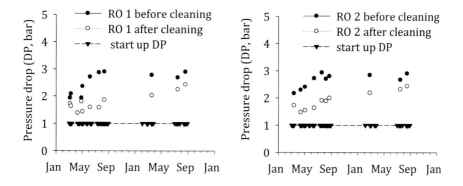

Figure 3.11: Historical (2015-2016) operational data for pressure drop (ΔP) in RO before and after CIP. The operational data was collected from the plant and processed.

3.3.2.5 Analysis of BRP method

The BRP method using flow cytometry was found fast (2-3 days) to achieve the maximum bacterial regrowth compared to the conventional plate counting which takes about 12-14 days. The increased inoculum density to 10^4 cells/mL from 50 - 500 CFU/mL as recommended by (Van der Kooij et al., 1982) and an incubation temperature of 30 °C might have helped in reaching maximum bacterial regrowth (N_{max}) within 1-3 days. One of the advantages of BRP method is that it can also be used to convert the net bacterial regrowth to the equivalent carbon concentration using the calibration curve. Accordingly, the detection limit of the method was found to be approximately 9.3 µg-$C_{glucose}$/L. The comparisons of BRP methods of this study with the current AOC methods are presented in Table 3.5. The most of the recent AOC methods have been developed for freshwater samples, and little research has been done on seawater samples (Table 3.5). Most recently, Jeong et al. (2013) and Weinrich et al. (2011) developed a method to determine AOC in seawater samples based on luminescence using the nutritionally diverse and naturally bioluminescent marine organisms *V. fischeri and V. harveyi*, respectively. The method developed by Jeong et al, (2013) was fast (< 1 h) and has a low limit of detection (0.1 µg-$C_{glucose}$/L). Moreover, the disadvantages of the method were;

i) Blank was subtracted to calculate the limit of detection. However, based on the provided calibration curves and bioluminescence signal of *V.fischeri*, the blank was approximately equivalent to 90 µg-$C_{glucose}$/L, which is very high. In the current BRP method, we managed to reduce the level of blank to 9.3 µg-$C_{glucose}$/L.

ii) Use of pure culture *V.fischeri* as inoculum which has the limitation of utilizing only certain biodegradable compounds (Servais et al., 1987)

Unlike other methods, the BRP method employed in this study used a natural bacterial consortium a as inoculum. It has the advantages that it can better adapt to metabolize organics present in seawater and thus provide more representative information regarding the biofouling potential of SWRO feed water.

Overall, the application of BRP and FCM is a rapid tool for assessing bacteria removal performance in a desalination plant, assess the efficiency of the pre-treatment system and indirectly provides a qualitative assessment of biofouling status of SWRO membranes (Dixon et al., 2012). However, whether or not the measured bacterial regrowth potential of SWRO feed water correlates with the occurrence of biofouling in SWRO membranes is still debatable and remains to be demonstrated in SWRO plants.

Table 3.5: Comparison of BRP with current AOC methods

Year	Author	Type of water	Bacterial elimination/removal	Substrate	AOC determination					Conversion /yield factor	Detection limit
					Bacterial Inoculation		Sample incubation		Bacterial enumeration		
					Strain	Volume	Temp (°C)	Time (d)			
1982	Van der Kooij	Freshwater	Pasteurization (60 °C), 30 min	Acetate	P-17, NOX	500 CFU/mL	15	7-9	Plate counting	4.6×10^6 (P17) 1.2×10^7 (NOX)	1 µg/L acetate C
1986	Werner and Hambsch	Freshwater	0.22 µm filtration	Acetate	NMC	5×10^4 TCN/mL	22	2-4	turbidity	2.3 ppm/L mg/L Acetate-C	10 µg/L acetate C
1993	Kaplan	Freshwater	Pasteurization (70 °C), 30 min	Acetate	P-17, NOX		25	5-9	Plate counting	N/A	
1993	Le Chevalier	Freshwater	Pasteurization (70°C), 30 min	Acetate	P-17, NOX	10^4 CFU/mL	22	2-4	ATP	N/A	N.A.
1999	Sathasivan	Freshwater	Sterilization 121°C, 15 min	Acetate	NMC	N.A	20	5	Total direct count	N/A	N.A.
2005	Hammes	Freshwater	0.22 µm filter	Acetate	NMC	10^4 CFU/mL	30	2-3	Flow cytometry	1×10^7	10 µg/L
2011, 2016	Weinrich	Seawater	Pasteurization (70°C), 30 min	Acetate	*V. harveyi*	10^3 CFU/mL	30	<1 d	Luminescence		<10 µg/L acetate C
2013	Jeong	Seawater	Sterilization (70 °C), 30 min & 0.22 µm filter	Glucose	*V. fischeri*	3×10^4 CFU/mL	25	<1 h	Luminescence		0.1 µg-C-glucose/L
2017	Dhakal	Seawater	0.22 µm Filtration	Glucose	NMC	10^4 cells/mL	30	1-3 days	Flow cytometry	4.5×10^6	9.3 µg-C-glucose/L

3.4   Conclusions

- The BRP method developed using FCM, and a natural consortium of bacteria as inoculum is a potential tool to evaluate the biofouling potential of seawater reverse osmosis (SWRO) feed water. The BRP method was relatively fast (2-3 days) compared to the conventional bioassays.

- The heating of chemical (NaCl) and bottle at 550 $^\circ$C for 6 hours during blank preparation was effective in lowering the detection limit of the BRP method to a level of 43,000 ± 12,000 cells/mL, which is equivalent to 9.3 ± 2.6 μg-$C_{glucose}$/L assuming a yield factor of 4.6 ×10^6 cells/μg-C for marine bacteria.

- The calibration of BRP method was developed in artificial and natural seawater using glucose as a standard substrate. Calibration results showed a good linearity (R^2 = 0.88 to 0.95) between a range of 0 – 2,000 μg-$C_{glucose}$/L.

- The method was applied to monitor the bacterial reduction potential of pre-treatment schemes of SWRO plants. Results showed that the bacterial regrowth potential reduction in SWRO feed water by DAF-UF as pre-treatment was 54 %, while it was 40 % with DMF-CF

- The measured bacterial regrowth potential values of SWRO feed water in DAF-UF-RO scheme was consistent with the frequency of chemical cleaning in the SWRO system. However, it is still debatable whether the measured bacterial regrowth potential of SWRO feed water directly correlate with the pressure drop increase in SWRO or not. More experiments are required to develop the relationship between the BRP and the pressure drop increase in SWRO plants.

3.5   References

Barth Jr, V.C., Cattani, F., Ferreira, C.A.S. and de Oliveira, S.D. (2012) Sodium chloride affects propidium monoazide action to distinguish viable cells. Analytical Biochemistry 428(2), 108-110.

Batté, M., Koudjonou, B., Laurent, P., Mathieu, L., Coallier, J. and Prévost, M. (2003) Biofilm responses to ageing and to a high phosphate load in a bench-scale drinking water system. Water Research 37(6), 1351-1361.

Csonka, L.N. (1989) Physiological and genetic responses of bacteria to osmotic stress. Microbiol Rev 53(1), 121-147.

Dixon, M.B., Qiu, T., Blaikie, M. and Pelekani, C. (2012) The application of the Bacterial Regrowth Potential method and Flow Cytometry for biofouling detection at the Penneshaw Desalination Plant in South Australia. Desalination 284, 245-252.

Filloux, E., Wang, J., Pidou, M., Gernjak, W. and Yuan, Z. (2015) Biofouling and scaling control of reverse osmosis membrane using one-step cleaning-potential of acidified nitrite solution as an agent. Journal of Membrane Science 495, 276-283.

Flemming, H.C., Schaule, G., Griebe, T., Schmitt, J. and Tamachkiarowa, A. (1997) Biofouling—the Achilles heel of membrane processes. Desalination 113(2–3), 215-225.

Gatza, E., Hammes, F. and Prest, E.I. (2013) Assessing water quality with the BD Accuri C6 Flow cytometer, BD Biosciences, White paper.

Greenlee, L.F., Lawler, D.F., Freeman, B.D., Marrot, B. and Moulin, P. (2009) Reverse osmosis desalination: Water sources, technology, and today's challenges. Water Research 43(9), 2317-2348.

Griebe, T. and Flemming, H.-C. (1998) Biocide-free antifouling strategy to protect RO membranes from biofouling. Desalination 118(1–3), 153-IN159.

Hambsch, B. and Werner, P. (1993) Control of bacterial regrowth in drinking water treatment plants and distribution systems. Water Supply 11, 299-308.

Hammes, F., Berger, C., Köster, O. and Egli, T. (2010) Assessing biological stability of drinking water without disinfectant residuals in a full-scale water supply system. Journal of Water Supply: Research and Technology - Aqua 59(1), 31-40.

Hammes, F., Berney, M., Wang, Y., Vital, M., Köster, O. and Egli, T. (2008) Flow-cytometric total bacterial cell counts as a descriptive microbiological parameter for drinking water treatment processes. Water Research 42(1–2), 269-277.

Hammes, F.A. and Egli, T. (2005) New Method for Assimilable Organic Carbon Determination Using Flow-Cytometric Enumeration and a Natural Microbial Consortium as Inoculum. Environmental Science Technology 39, 3289-3294.

Jacobson, J.D., Kennedy, M.D., Amy, G. and Schippers, J.C. (2009) Phosphate limitation in reverse osmosis: An option to control biofouling? Desalination and Water Treatment 5, 198-206.

Jeong, S., Naidu, G., Vigneswaran, S., Ma, C.H. and Rice, S.A. (2013) A rapid bioluminescence-based test of assimilable organic carbon for seawater. Desalination 317, 160-165.

Kaplan, L.A., Bott, T.L. and Reasoner, D.J. (1993) Evaluation and simplification of the assimilable organic carbon nutrient bioassay for bacteria growth in drinking water. Applied and environmental microbiology 59, 1532-1539.

Kim, D., Jung, S., Sohn, J., Kim, H. and Lee, S. (2009) Biocide application for controlling biofouling of SWRO membranes — an overview. Desalination 238(1–3), 43-52.

Le Chevalier, M.W., Shaw, N.E., Kaplan, L.A. and Bott, T.L. (1993) Development of a rapid assimilable organic carbon method for water. Applied and environmental microbiology 29(5), 1526-1531.

Matin, A., Khan, Z., Zaidi, S.M.J. and Boyce, M.C. (2011) Biofouling in reverse osmosis membranes for seawater desalination: Phenomena and prevention. Desalination 281, 1-16.

Passow, U. (2002) Transparent exopolymer particles (TEP) in aquatic environments. Progress in Oceanography 55(3–4), 287-333.

Prest, E.I., Hammes, F., Kötzsch, S., van Loosdrecht, M.C.M. and Vrouwenvelder, J.S. (2016) A systematic approach for the assessment of bacterial growth-controlling factors linked to biological stability of drinking water in distribution systems. Water Science and Technology: Water Supply 16(4), 865-880.

Prest, E.I., Hammes, F., Kötzsch, S., van Loosdrecht, M.C.M. and Vrouwenvelder, J.S. (2013) Monitoring microbiological changes in drinking water systems using a fast and reproducible flow cytometric method. Water Research 47(19), 7131-7142.

Quek, S.B., Cheng, L. and Cord-Ruwisch, R. (2015) Microbial fuel cell biosensor for rapid assessment of assimilable organic carbon under marine conditions. Water Research 77, 64-71.

Ross, P.S., Hammes, F., Dignum, M., Magic-Knezev, A., Hambsch, B. and Rietveld, L.C. (2013) A comparative study of three different assimilable organic carbon (AOC) methods: results of a round-robin test. Water Science and Technology: Water Supply 13(4), 1024-1033.

Sathasivan, A. and Ohgaki, S. (1999) Application of new bacterial regrowth potential method for water distribution system – a clear evidence of phosphorus limitation. Water Research 33(1), 137-144.

Servais, P., Billen, G. and Hascoët, M.-C. (1987) Determination of the biodegradable fraction of dissolved organic matter in waters. Water Research 21(4), 445-450.

Stiefel, P., Schmidt-Emrich, S., Maniura-Weber, K. and Ren, Q. (2015) Critical aspects of using bacterial cell viability assays with the fluorophores SYTO9 and propidium iodide. BMC Microbiology 15, 36.

Tsueng, G. and Lam, K.S. (2010) A preliminary investigation on the growth requirement for monovalent cations, divalent cations and medium ionic strength of marine actinomycete Salinispora. Applied Microbiology and Biotechnology 86(5), 1525-1534.

Unemoto, T., Tsuruoka, T. and Hayashi, M. (1973) Role of Na+ and K+ in preventing lysis of a slightly halophilic Vibrio alginolyticus. Canadian Journal of Microbiology 19(5), 563-571.

Vaccaro, R.F., Hicks, S.E., Jannasch, H.W. and Carey, F.G. (1968) The occurrence and role of glucose in seawater. Limnology and Oceanography 13(2), 356-360.

Van der Kooij, D., Visser, A. and Hijnen, W.A.M. (1982) Determination of easily assimilable organic carbon in drinking water. Journal of the American Water Works Association. 74, 540-545.

Van der Merwe, R., Hammes, F., Lattemann, S. and Amy, G. (2014) Flow cytometric assessment of microbial abundance in the near-field area of seawater reverse osmosis concentrate discharge. Desalination 343, 208-216.

Villacorte, L.O. (2014) Algal blooms and membrane based desalination technology, Ph.D. thesis UNESCO-IHE/TU Delft

Vital, M., Dignum, M., Magic-Knezev, A., Ross, P., Rietveld, L. and Hammes, F. (2012) Flow cytometry and adenosine tri-phosphate analysis: Alternative possibilities to evaluate major bacteriological changes in drinking water treatment and distribution systems. Water Research 46(15), 4665-4676.

Vrouwenvelder, J.S., Beyer, F., Dahmani, K., Hasan, N., Galjaard, G., Kruithof, J.C. and Van Loosdrecht, M.C.M. (2010) Phosphate limitation to control biofouling. Water Research 44(11), 3454-3466.

Vrouwenvelder, J.S., Van Paassen, J.A.M., Wessels, L.P., Van Dam, A.F. and Bakker, S.M. (2006) The membrane fouling simulator: a practical tool for fouling prediction and control. Journal of Membrane Science 281 (1-2), 316-324.

Weinrich, L., Haas, C.N. and LeChevallier, M.W. (2013) Recent advances in measuring and modeling reverse osmosis membrane fouling in seawater desalination: a review. Journal of Water Reuse and Desalination 3(2), 85-101.

Weinrich, L., LeChevallier, M. and Haas, C.N. (2016) Contribution of assimilable organic carbon to biological fouling in seawater reverse osmosis membrane treatment. Water Research 101, 203-213.

Weinrich, L.A., Schneider, O.D. and LeChevallier, M.W. (2011) Bioluminescence-Based Method for Measuring Assimilable Organic Carbon in Pretreatment Water for Reverse Osmosis Membrane Desalination. Applied and environmental microbiology 77, 1148-1150.

Werner, P. and Hambsch, B. (1986) Investigations on the growth of bacteria in drinking water. Water Supply 4 227–232.

Withers, N. and Werner, P. (1998) Bacterial regrowth potential:quantitative measure by acetate carbon equivalents. Water 25 (5), 19-23.

Zipper, H., Brunner, H., Bernhagen, J. and Vitzthum, F. (2004) Investigations on DNA intercalation and surface binding by SYBR Green I, its structure determination and methodological implications. Nucleic Acids Research 32(12), e103-e103.

Supplementary – S1

Supplementary Table S.3.1: Inorganic ion composition of model artificial seawater (ASW)

Inorganic Ions	Concentration (g/L)
Chlorine (Cl^-)	18.85
Sodium (Na^+)	10.75
Sulphate (SO_4^{-2})	2.69
Magnesium (Mg^{2+})	1.17
Calcium (Ca^{2+})	0.30
Potassium (K^+)	0.38
Hydrogen Carbonate (HCO_3^-)	0.15
Total dissolved solids (TDS)	34.29

Supplementary – S2

Supplementary Table S.3.2: Live bacterial cell enumeration by FCM for serially diluted seawater samples

% of dilution	Live bacterial cells Cells/mL [a]	Percentage deviation
1%	6,333 ± 577	9.1 %
2%	12,000 ± 675	4.8 %
4%	29,667 ± 460	1.9 %
10%	57,333 ± 385	1.0 %
20%	116,333 ± 1,528	1.3 %
25%	140,000 ± 1,155	0.8 %
50%	278,333 ± 1,732	0.6 %
100%	541,333 ± 7,937	1.5 %

[a]Values are average ± standard deviation; n = 3

Supplementary – S3 Natural consortium bacterial growth curve using acetate and glucose as substrate

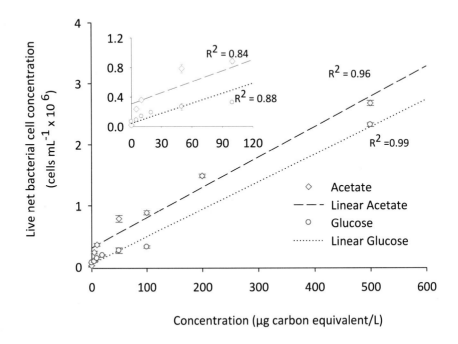

Supplementary Figure S.3.1: Natural consortium bacterial growth curve using acetate and glucose as substrate. Error bars indicate standard deviations of triplicate measurements. The inset graph provides the reading in the lower range of concentration

Supplementary – S4 Flow cytometry calibration and validation

Calibration of FCM (Partec calibration beads, 3 µm (daily control)

The accuracy of the FCM measurement was checked using the Partec calibration beads before any daily measurement.

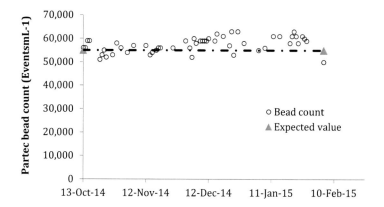

Supplementary Figure S.3.2: Partec calibration beads, 3 µm (daily control beads provided by the FCM bead provider); measurement was carried out on each day's operation of the flow cytometer.

Supplementary Figure S.3.2 shows that the average beads counts during the study period were 57,519 Events/mL with 5.9 % deviation from the expected value. The bead counts were within the expected count ranges provided by the guideline (50,000 to 60,000 events/mL).

8 and 6 - Peak calibration beads

The FCM laser position was checked occasionally using 6-and 8-peaks calibration beads checks. Supplementary Figure S. 3.3 and 3.4 indicate the expected number of peaks recorded in each respective graph during the calibration process.

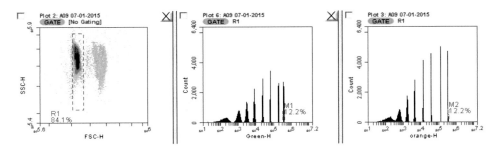

Supplementary Figure S.3.3: 8-peaks calibration beads check

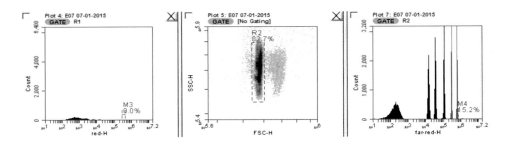

Supplementary Figure S.3.4: 6-peaks calibration beads check

Supplementary – S5: Comparison of seawater gating strategy

The proposed gating in this study was compared with the other existing gating such as seawater gating propose by NIOZ and freshwater gating proposed by Prest et al. (2013)

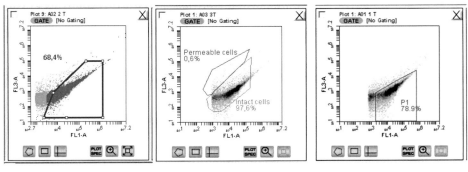

a) New proposed seawater   b) Seawater gating proposed by   c) Freshwater gating proposed
   gating   NIOZ   by Prest (Prest et al., 2013)

Supplementary Figure S.3.5: Comparison of seawater gating strategies

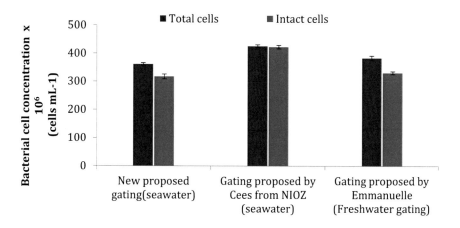

Supplementary Figure S.3.6: Comparison of total cell and intact cell counting using different seawater gating strategies

Fouling of ultrafiltration membranes by organic matter generated by four marine algal species

Contents

This chapter is based on:

Dhakal, N., Salinas Rodriguez, S.G., Ouda, A., Schippers, J.C. and Kennedy, M.D. (2017), Fouling of ultrafiltration membranes by organic matter generated by four marine algal species. *Submitted to Journal of Membrane Science*

Abstract

Controlling fouling in seawater reverse osmosis and ultrafiltration systems applied for pre-treatment is a major challenge, in particular during algal blooms. This study aims to investigate the fouling potential and fouling behavior of algae and released organic matter in ultrafiltration membranes. For this purpose, four marine algae were cultivated namely: Chaetoceros affinis (Ch), Rhodomonas balthica (Rb), Tetraselmis suecica (Te), and Phaeocystis globulosa (Ph). During the growth and stationary/decline phase, the algal cell density, chlorophyll-a, biopolymer, transparent exopolymer particles (TEP) concentration and MFI-UF$_{10kDa}$ (membrane fouling potential) were measured. Fouling experiments were executed with capillary ultrafiltration, filtration inside to outside, and backwashable and non- backwashable fouling was monitored.

During the growth, stationary/decline phase of the algal species remarkable differences were observed in the production of biopolymers, TEP and MFI-UF$_{10kDa}$. Membrane fouling potential (MFI-UF 10 kDa) was linearly related to algal cell density and chlorophyll-a concentration, biopolymer concentration, TEP, during the growth phase of the algal species. After the growth phase, the relationship between MFI-UF$_{10kDa}$ and algal cell density and chlorophyll-a concentration did not continue. In experiments with capillary ultrafiltration, membranes (150 kDa) fed with water having 0.5 mg-biopolymer-C/L back washable fouling coincided with the MFI-UF$_{150kDa}$ TEP for Rh, Te, and Ph. Back washable fouling of Ch deviated and was substantially higher. The non-back washable fouling of the ultrafiltration membranes varied strongly with the type of algal species and coincided with MFI-UF$_{150kDa}$ and TEP concentration. Rh demonstrated the highest and Ph the lowest non-back washable fouling (at a level of 0.5 mg-biopolymer-C/L in the feed water. This non-backwashable fouling is attributed to polysaccharides (stretching - OH) and sugar ester group (stretching S=O) present in the AOM.

Overall, this study highlights the importance of having better monitoring methods, types of bloom-forming algae and characteristics of AOM they released to predict the non-backwashable fouling in dead-end UF membranes.

Keywords: *Marine algae, algal organic matter, fouling ultrafiltration membranes, MFI-UF, TEP, biopolymers.*

4.1   Introduction

Ultrafiltration (UF) is widely applied as pre-treatment in seawater reverse osmosis (SWRO) desalination plants (Bonnélye et al., 2008, Villacorte et al., 2015c, Voutchkov, 2010). The rationale behind the rapidly expanding application of ultrafiltration at the expense of conventional pre-treatment systems as illustrated in Figure 4.1 is that it is very effective in removing the particulate, colloidal matter ensuring permanently low silt density index (SDI) values (Tian et al., 2013). However, controlling fouling is a major challenge in particular during algal blooms. Algal bloom, or a so-called "population explosion" of algae, is still considering a leading cause of operational problems in both SWRO and the pre-treatment systems (Caron et al., 2010, Villacorte et al., 2015c). A notable example is the severe algal bloom caused by the dinoflagellate *Cochlodinium polykrikoides* in the Gulf region of the Middle East in 2008 - 2009 (Richlen et al., 2010). The criterion to define algal blooms are algal cell concentration > 1million cells/L and chlorophyll-a concentration >10 μg/L)(http://www.waterman.hku.hk, 2016). The algal bloom event of 2008-2009 in the Middle East forced several desalination plants to shut down the operations mainly due to clogging of pre-treatment systems used, i.e., granular media filters (GMF) (Pankratz, 2008, Reddy, 2009). The consequences of rapid clogging of the media filters are less water production due to frequent backwashing of the filters, and poor quality of SWRO feed water (silt density index > 5) which could irreversibly foul SWRO membrane modules. Since then, the application of UF as pre-treatment is gaining a market in seawater desalination industry.

The impact of algal blooms in seawater on the performance of ultrafiltration has been reported by Schurer et al., (2013). An ultrafiltration pilot study on North Seawater showed that severe non-backwashable fouling could not be controlled without the use of pre-treatment namely in-line coagulation (Schurer et al., 2013). Although numerous advantages of this technology, algal blooms and the development of non-backwashable fouling resistance is a major threat to the operation of UF membranes. During an algal bloom cell density and the algal released organic matter are responsible for membrane fouling (Caron et al., 2010). Elevated levels of algal organic matter (AOM) is responsible for the need of frequent hydraulic backwashing and chemical enhanced backwashing (CEB). Fouling compounds that are tightly bound to the UF membrane are referred to hydraulically irreversible membrane fouling, and such fouling increased the frequency of

CEB (Peldszus et al., 2011). To overcome this problem, conventional techniques as a pre-treatment are gradually introduced, e.g., inline coagulation or dissolved air floatation (DAF) upfront of ultrafiltration. In general pre-treatment with in-line coagulant dosing in ultrafiltration, feed water is expected to be able to control severe fouling (Schurer et al., 2013). However, environmental and sustainability considerations urge for minimizing or avoiding the use of chemicals. This study focuses on investigating fouling, caused by algal growth or bloom, of ultrafiltration membranes operating without pre-treatment making use of a coagulant.

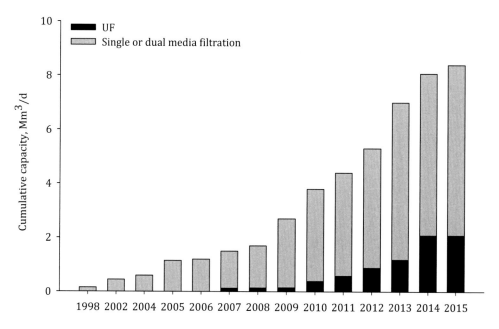

Figure 4.1: Cumulative capacity of the forty largest seawater RO desalination plants per type of pre-treatment as in 2016 (Arevalo et al., 2017)

It has been shown that accumulation of algal organic matter (AOM) during algal bloom is the leading cause of membrane fouling, rather than the algae cells themselves (Ladner et al., 2010, Qu et al., 2012, Schurer et al., 2012, Villacorte et al., 2015c). The AOM consists of polysaccharides, proteins, lipids, nucleic acids, and other dissolved organic substances (Fogg, 1983, Myklestad, 1995). A fraction of AOM are transparent exopolymer particles (TEP)) (Villacorte et al., 2013) which are very sticky polysaccharides and glycoproteins (Passow et al., 1995). The TEP-like matter is expected to cause severe fouling in UF and

RO systems and may initiate biological fouling in RO systems (Berman et al., 2005, Berman et al., 2011, Kennedy et al., 2009). To further develop strategies, preferably environmentally friendly, to control membrane fouling, the characteristics of the foulants need to be identified and monitored. Parameters which potentially can be used are algal cell density, chlorophyll-a concentration, algal biopolymer, transparent exopolymeric particles (TEP_{10kDa}), membrane fouling potential ($MFI-UF_{10ka}$).

This study aims to investigate the fouling potential of marine bloom-forming algae and the AOM released in ultrafiltration membrane. Furthermore, the study also aims to investigate the relationship between the fouling parameters and the backwashable and non-backwashable fouling rate in UF membranes. For this purpose, four marine algal species were selected for cultivating in the laboratory. Selected were algae species commonly growing in three different seas and one ocean namely: Baltic Sea, Mediterranean Sea, North Sea and the Pacific Ocean. These algal species are respectively; *Rhodomonas balthica (Rh), Chaetoceros affinis (Ch), Phaeocystis globulosa (Ph)*, and *Tetraselmis suecica (Te)*. The specific objectives are:

i.   To monitor parameters potentially related to fouling potential of four different algal species at different stages of their growth, generated in the four different solutions

ii. To investigate the relation between the fouling potential (as measured by $MFI-UF_{10kDa}$) in the four different solutions with the parameters: algal cell density, chlorophyll-a, biopolymer, and transparent exopolymeric particles (TEP_{10kDa}).

iii. To investigate the relation between the backwashable and non-backwashable fouling rate in ultrafiltration experiments, executed with diluted solutions, with measured fouling parameters.

iv. To identify the nature of non-backwashable fouling compounds on ultrafiltration membrane using Fourier Transform InfraRed (FTIR) spectroscopy.

4.2   Material and methods

4.2.1 Algal cultures

Four strains of marine bloom-forming algal species were selected for this study, *Chaetoceros affinis* (CCAP 1010/27), *Rhodomonas balthica* (NIVA 5/91), *Tetraselmis suecica* (CCAP 66/22), and *Phaeocystis globulosa* (CCY 0801). The four strains were inoculated in natural North Sea water (TDS 34 g/L, pH = 8 ± 0.3). Among them, the two strains of *Chaetoceros affinis* and *Phaeocystis globulosa* were spiked with nutrients and trace elements based on the F/2 with Silicate and L1 without silicate medium, respectively, while strains of *Rhodomonas balthica* and *Tetraselmis suecica* were grown in a F/2 medium. For this purpose, North Sea water was pre-filtered with 2 μm glass filter and autoclaved (121 °C, 20 minutes) before use. All algal cultures except *Phaeocystis globulosa* were exposed to a continuous mercury fluorescent light. With *Phaeocystis globulosa*, the culture was exposed to 16 h light / 8 h dark, as this was expected to be optimal for its growth. Aeration was employed for mixing.

4.2.2 Cell density and chlorophyll-a

The algal-cell concentration in batch cultures was monitored by sampling every two days and counting the cells using Haemacytometer (Burker-Turk counting chamber) slides and a light Nikon microscope (Olympus BX51). The flagellate types of algal species were immobilized with Lugol's iodine solution before counting. Samples were also collected to measure the chlorophyll-a concentration according to the Dutch standard NEN 6520 protocol. In this method, the algae solution was filtered (GF 6 filter) and extracted with ethanol. The absorbance of a sample before and after acidification (0.4 M HCl) was measured at wavelengths of 665 and 750 nm using a spectrophotometer. The difference in absorbance of the solution is the measure of chlorophyll-a content.

Additional samples were collected for liquid chromatography organic carbon detection (see Section 4.2.4.1), transparent exopolymer particles (TEP_{10kDa}) (see Section 4.2.4.2), and modified fouling index ($MFI\text{-}UF_{10kDa}$) measurements (see Section 4.2.4.5). The samples were collected every two days until the stationary/decline phase of the algal growth.

4.2.3 Extraction and characterization of algal organic matter (AOM)

Before using water for further testing, samples were first allowed to settle for 24 hours to remove the larger algal cells. Subsequently, the supernatant containing AOM was separated and filtered through a 5 µm polycarbonate filter (Nuclepore PC membranes, Whatman) with < 0.2 bars of vacuum. The filtered samples were analyzed using:

- Liquid chromatography - organic carbon detection (LC-OCD) (see section 4.2.4.1)
- Transparent exopolymer particles (TEP$_{10kDa}$) (see Section 4.2.4.2)
- Fluorescence excitation-emission matrix (F-EEM) spectroscopy (see section 4.2.4.3)
- Fourier transform infrared (FTIR) spectroscopy (see section 4.2.4.4)
- Modified fouling index (MFI-UF$_{10kDa}$) (see Section 4.2.4.5),

The extracted AOM solution was used as a feed solution during the ultrafiltration experiments (see section 4.2.5) and compared the backwashable and non-backwashable fouling rate development.

4.2.4 Characterization techniques

4.2.4.1 Liquid chromatography - organic carbon detection (LC-OCD)

AOM samples extracted from algal cultures were analyzed at Wetsus, Leeuwarden, the Netherlands using liquid chromatography organic carbon detection (LC-OCD). The LC-OCD is based on the size-exclusion chromatography, and is equipped with an organic carbon detector (OCD), organic nitrogen detector (OND), and UV$_{254}$ detector. The LC-OCD analysis determined the organic carbon concentrations of biopolymers, humic substances, building blocks, low molecular weight (LMW) acids, and neutrals. Before LC-OCD analyses, all samples were pre-filtered through 0.45 µm Millipore filters. The measurement and analysis of the samples were performed according to the protocol described by (Huber et al., 2011).

4.2.4.2 Transparent exopolymer particles (TEP$_{10kDa}$) measurement

TEP$_{10kDa}$ were measured according to the protocol described by (Villacorte et al., 2015a). In short, the water sample was filtered through a 10 kDa regenerated cellulose Millipore membrane filter, and the filtered volume (V$_f$) was measured. The retained TEP concentration on the filter paper was re-suspended in MilliQ water (V$_r$ = 10 mL) and stained with alcian blue (AB) that was pre-filtered through 0.05 µm polycarbonate filter. The TEP-AB solution (4 mL) was again filtered through a 0.1 µm polycarbonate filter at <0.2 bar vacuum pressure. The absorbance (A$_e$) of the filtered sample and blank (A$_b$) were measured using a spectrophotometer at a wavelength of 610 nm. The TEP$_{10kDa}$ concentration in mg Xanthan equivalent per liter was calculated using equation 4.1 and 4.2.

$$TEP_{10\,kDa} = f_{610nm\frac{V_r}{V_f}(A_e-A_b)} \qquad \text{[Equation 4.1]}$$

$$f_{610nm} = \frac{1}{m_{610nm}} \qquad \text{[Equation 4.2]}$$

Where:

f_{610nm} : Calibration factor [(mg X$_{eq}$/L)/ (abs/cm)];

m_{610nm} : Slope of the calibration line [(abs/cm)/ (mg X$_{eq}$/L)]

4.2.4.3 Fluorescence excitation-emission matrix (FEEM) spectroscopy

The fluorescence emitting organic substances in AOM samples were measured using a FluoroMax-3 spectrophotometer (Horiba Jobin Yvon, Inc., USA) with a 150 W ozone-free xenon arc lamp to enable excitation. The AOM solutions were scanned over the excitation wavelength range from 240 to 450 nm, and an emission wavelength range of 290 to 500 nm to produce a three-dimensional matrix. Before F-EEM analysis, the dissolved organic carbon (DOC) of AOM samples was measured using Shimadzu TOC, and diluted if necessary with MilliQ water to have a DOC concentration of approximately 1 mg/L. Excitation and emission matrices were analyzed using MatLab R 2011a, and the results were interpreted as described by (Leenheer et al., 2003).

4.2.4.4 Fourier transform infrared (FTIR) spectroscopy

FTIR spectroscopy was applied to identify the functional groups present on the AOM fouled ultrafiltration membranes before and after the cleaning of the membrane. The

AOM samples were first filtered using a flat sheet Millipore ultrafiltration membrane (molecular weight cut-off, 10 kDa) at a constant flux of 60 L/m²/h. The AOM-fouled membranes were analyzed using a PerkinElmer ATR-FTIR Spectrum-100 instrument at the Aerospace Engineering Laboratory of the Delft University of Technology. The AOM-fouled membrane was placed directly on the ATR crystal and held in place with a loading screw. An average of 16 scans between 4,000 and 400 cm^{-1} wavenumbers was recorded. The peak-picking feature of the spectrum analysis software was used to identify major peaks of interest.

The AOM-fouled membrane (after FTIR test) was placed in a clean disposable plastic container (40 mL) containing 10 mL of synthetic seawater. The samples were tightly covered and vortexed (Heidolph REAX 2000) for 10 seconds and sonicated (Branson 2510E-MT) for 90 minutes at 42 kHz to remove loosely bound AOM. The FTIR of the AOM-fouled membrane after sonication was performed and compared with the FTIR results from before sonication.

4.2.4.5 Modified fouling index (MFI - UF)

The Modified Fouling Index (MFI-UF$_{10kDa}$) was measured at constant flux through a membrane having pores of 10 kDa according to the protocol developed by (Boerlage et al., 2000) and modified by (Salinas-Rodriguez et al., 2015). Membranes having pores of 10 kDa were used because it is indicated or suggested that particles down to 10 kDa are likely responsible for particulate fouling of RO membranes. The water sample was filtered at a constant flux of 60 L/m²/h using syringe pumps, and filter holder (Schleicher & Schuell). The trans-membrane pressure (ΔP) development over time was recorded using a pressure sensor. The data obtained for ΔP were plotted, and the minimum slope was calculated to determine the fouling index (I). MFI-UF was then calculated by normalizing the fouling index with standard reference values as proposed by Schippers and Verdouw in 1980 (Schippers et al., 1980) as equation 4.3.

$$MFI - UF = \frac{\eta_{20^o C}\, I}{2\Delta P_o A_o^2}$$
[Equation 4.3]

Where $\eta_{20^o C}$ is the water viscosity at 20 °C, ΔP represents a standard feed pressure (2 bars) and A_o is a standard membrane area of 13.8 x 10^{-4} m².

4.2.4.6 Silt density index (SDI)

Silt density index is determined by measuring the rate of plugging of a 0.45 µm membrane filter at 2 bars according to the standard ASTM protocol. The measurement is done as follows.

- time t_1 required to filter the first 500 mL is determined
- 15 minutes (t_f) after the start of this measurement, time t_2 needed to filter another 500 mL is determined
- SDI is calculated using the equation 4.4

$$SDI = \frac{100\,\%}{t_f}\left(1 - \frac{t_1}{t_2}\right) = \frac{\%\,P}{t_f} \qquad\qquad \text{[Equation 4.4]}$$

The shorter time (t_f) has to be considered such as 10, 5 or 3 minutes if the plugging ratio (% P) exceeds 75 %. The volume of filtered water sample is proportional to the diameter of the filter used.

4.2.4.7 MFI$_{0.45}$

The Modified fouling index (MFI$_{0.45}$) was developed by Schippers and Verdouw (1980) and is based on the cake filtration model (Schippers et al., 1980). For determination of MFI$_{0.45}$, the flow through the membrane filter is measured as a function of time.

$$\frac{t}{V} = \frac{\mu.R_M}{dP.A} + \frac{\mu.I}{2.\Delta P.A_M^2}V \qquad\qquad \text{[Equation 4.5]}$$

Where

V	=filtrate volume (L or m³)
T	=time (s)
A_M	=membrane area (m²)
dP	=applied pressure (Pa)
µ	=water viscosity (Pa.S)
R_M	=clean membrane resistance (m⁻¹)
I	=fouling potential (m⁻²)

The MFI$_{0.45}$ is calculated from the slope of t/V versus V graph and is corrected for pressure and temperature.

4.2.5 Preparation of membrane pen modules

Hollow-fiber polyethersulphone (PES) membranes (molecular weight cut-off, 150 kDa) obtained from Pentair X-Flow were used. The preparation of membrane pen modules was according to the protocol of (Tabatabai et al., 2014). In short, membrane pen modules were fabricated by potting four hollow fiber membranes with an internal diameter of 0.8 mm in 30 cm transparent 8 mm outer diameter polyethylene tubing (Festo, Germany). Membrane pen modules were potted using polyurethane glue (Bison, the Netherlands). The effective surface area of the membrane was 30 cm^2 ± 2 %. A new membrane pen module was used for each filtration experiment, and the prepared modules were soaked in 40 °C water for 24 hours. All the entrapped air inside and outside the fibers was removed before filtration experiments, and the modules were flushed with MilliQ water for approximately 40 minutes prior to the experiments.

4.2.6 Ultrafiltration experiments

The experiments were performed using a bench-scale filtration setup as shown in Figure 4.2 at a room temperature (20 °C ± 2 °C). Filtration was done in inside to outside mode at a flux of 80 L/m^2/h. A 20 minutes filtration cycle, followed by 45 seconds of backwashing with UF permeate at a flux of 200 L/m^2/h was employed in the tests. The transmembrane pressure (TMP) development in each filtration cycle was recorded using a pressure sensor (Cerabar PMC 55, Endress and Hauser, Switzerland). The operational pressure range of Cerabar PMC 55 was 0 - 4 bars with a maximum deviation of 0.04 %. The modem, FAX 195 Hart (Endress and Hauser, Switzerland) was connected to logged data on a computer.

The four different AOM solutions originating from the four algal species were fed to the UF filtration membranes. AOM solutions filtered through 5 μm filters and diluted with 2 μm filtered and autoclaved (121 ₀C, 20 minutes) North Sea water to have the final biopolymer concentration of 0.5 ± 0.15mg-C/L as a feed solution. Multi ultrafiltration cycles were performed for each AOM solution.

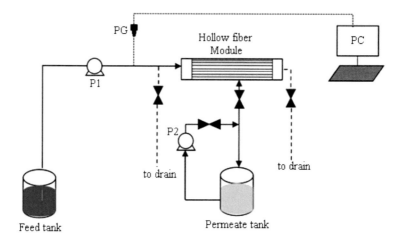

Figure 4.2: Laboratory scale filtration setup.

4.2.6.1 Backwashable and non-backwashable fouling rate calculation

The backwashable and non-backwashable fouling rates were calculated by plotting transmembrane pressure development over filtration time (Figure 4.3).The slope of the transmembrane pressure

- at the beginning of each filtration cycle (after hydraulic backwashing) is non-backwashable fouling (f_{nBW}).
- at the end of each filtration cycle (before hydraulic backwashing) is the total fouling rate (f_T).

The backwashable fouling rate is then calculated using Equation 4.4.

$$f_T = f_{BW} + f_{nBW}$$ [Equation 4.4]

Where, f_T, f_{BW}, and f_{nBW} represent the total, backwashable, and non-backwashable fouling rate (bar/h), respectively.

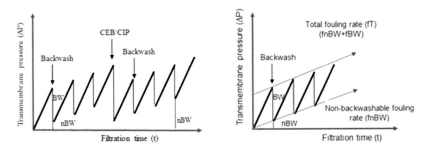

Figure 4.3: Fouling development in constant flux dead-end UF systems

4.2.6.2 Characterization of feed and permeate of UF membranes

The biopolymer concentration present in feed and permeate of UF membranes, collected after the filtration with four different AOM solutions, were analyzed using LC-OCD analysis as discussed in section 4.2.4.1.

4.2.7 Theoretical calculation of fouling potential of algal suspension

The fouling potential of algal suspension without AOM was calculated using the model Equation 5 as described by (Villacorte, 2014).

$$MFI = \frac{15\pi d_P C_P}{\varphi^2} \frac{(1-\epsilon)}{\epsilon^3} \frac{\eta_{20^oC}}{\Delta P_o A_o^2}$$
[Equation 4.5]

Where:

ε   =cake porosity

φ   =sphericity of the particles

Cp   =particle concentration (count/mL)

Dp   =diameter of particles comprising the cake

η_{20^oC} =water viscosity at 20 °C (0.001003 Pa.s)

ΔP_o =reference feed pressure (2 bar)

A_o =reference membrane area (13.8 x 10^{-4} m^2)

4.3 Results and discussion

4.3.1 Characteristics of cultured algae

The typical features of the four algal species are summarized in Table 4.1. The development in the algal cell density, MFI-UF$_{10kDa}$, chlorophyll-a, TEP$_{10kDa}$, and biopolymer concentration in each culture was monitored as shown in Figure 4.4. These parameters will be discussed below.

Table 4.1: Typical characteristics of the four algal species investigated this study

Characteristics	*Rhodomonas balthica (Rh)*	*Tetraselmis suecica (Te)*	*Chaetoceros affinis (Ch)*	*Phaeocystis globulosa (ph)*
Strain	NIVA 5/91	CCAP 66/22	CCAP 1010/27	CCY0801
Type	Cryptophyta	Flagellates	Diatom	Haptophytes
Geometric shape	Cone + half sphere	Ovoid, slightly flattened	Oval cylinder	Spherical
Size	8 - 12 µm	9 - 11 µm	8 - 10 µm	3 - 8 µm
Colour	Reddish	Green	Golden brown	White
Strain origin	Baltic Sea	Pacific Ocean	Mediterranean Sea	North Sea

Source: (http://nordicmicroalgae.org/taxon/Rhodomonas%20baltica, 2017, http://www.ccap.ac.uk/ccap-search.php, 2017)

4.3.2 Batch algal culture monitoring and fouling potential

The development in the algal cell density, MFI-UF$_{10kDa}$, chlorophyll-a, TEP$_{10kDa}$, and biopolymer concentration of four laboratory grown marine algae species were monitored as shown in Figure 4.4. These parameters will be discussed below.

4.3.2.1 Algal cell density and chlorophyll-a

In all cases, a lag phase of 1-2 days was followed by an (exponential) growth phase, and a stationary/ declining phase was observed. The increase in the algal cell density during the growth phase is attributed to the sufficiently high level of nutrients in the solution. After this phase, a stationary/decline phase was observed.

The concentration of algal cell density and chlorophyll-a were, as expected, much higher than the levels which are indicating an algal bloom. Algal blooms are considered to occur when algal cell density exceeds 1000 cells/mL (Villacorte et al., 2015d), and chlorophyll-a concentration exceeds 10 µg/L (http://www.waterman.hku.hk, 2016). Consequently, the solutions had to be diluted to a level, which is more likely to occur in practice. In this study, a concentration of 0.5 mg biopolymer per L was used, because this level occurred regularly in the North Sea during algal blooms.

4.3.2.2 TEP$_{10kDa}$ and biopolymer concentration

Remarkable differences were observed in the concentration of TEP$_{10kDa}$ produced by the four species of algae. In all cases, a very low concentration of TEP was measured in the lag phase (2-3days) followed by a rapid increase during the growth and stationary/decline phase. According to the results, the *Ch* culture produced the highest concentration of TEP$_{10kDa}$ (44 mgXeq/L), which was about 2-3 times higher than the *Rh, Te,* and *Ph* culture. The TEP concentration is, as expected, much higher than during algal blooms as it is higher than 1 mgXeq/L which was observed by Villacorte in the North Sea. (Passow, 2002). In three cases, namely *Ch, Rh and Te* the biopolymer and TEP concentration and MFI-UF$_{10kDa}$ continued increasing after arriving at the stationary phase. This observation is attributed to the release TEP from dead algal cells.

Likewise, remarkable differences were also observed among the four algal species regarding biopolymer concentration. The *Ch* culture produced the highest biopolymer concentration (46 mg-C/L), which was approximately 6, 8 and 15 times higher than cultures of *Te, Ph,* and *Rh*, respectively. The biopolymer concentration increased even during the stationary/death phase, which could be due to cell lysis and release of intracellular organic matter in the solution (Merle et al., 2016, Voutchkov, 2010). Villacorte et al. (2015) and Henderson et al. (2008) also described a similar phenomenon for fresh and marine algal species (Henderson et al., 2008, Villacorte et al., 2015b).

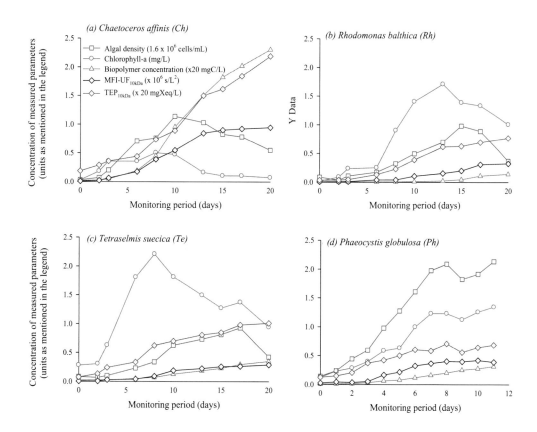

Figure 4.4: Development of algal cell density, MFI - UF, TEP, Chlorophyll a, Biopolymer concentration of batch cultures of (a) Rhodomonas balthica, (b) Chaetoceros affinis (c) Tetraselmis suecica, and (d) Phaeocystis globulosa.

4.3.2.3 Modified fouling index (MFI-UF$_{10\ kDa}$)

Among the four algal species, the culture of *Ch* recorded the highest MFI - UF value (946,000 s/L^2), which was approximately 2-3 times higher than the cultures of *Rh, Te,* and *Ph.*

The maximum values of the measured parameters for the four algal species are presented in Table 4.2.

Table 4.2: Maximum value of algal cell counts, chlorophyll-a, biopolymer, TEP concentration and MFI-UF of four marine algal species

Water quality parameters	Ch	Rh	Te	Ph
Algal cell (x 10^6 cells/mL)	1.8	1.6	1.5	3.4
Chlorophyll-a (µg/L)	503	1,700	2,200	1,300
Biopolymers (mg-C/L)	46	3	7	6
TEP$_{10kDa}$ (mgXeq/L)	44	15	20	14
MFI-UF$_{10kDa}$ (× 1,000) s/L^2	946	324	290	417

The specific MFI-UF$_{10kDa}$ is given in Table 4.3. These values show remarkable variations in particular for chlorophyll and biopolymers.

Table 4.3: Specific MFI-UF$_{10kDa}$

MFI-UF$_{10kDa}$ (x1000 s/L^2)	Ch	Rh	Te	Ph
per 10^6 algal cells	526	203	193	123
per µg/L chlorophyll-a	1,89	0,19	0,13	0,32
per mg-C/L Biopolymer	21	108	41	70
per mgXeq/L TEP$_{10kDa}$	21	22	15	30

A linear regression analysis showed a relationship between the MFI-UF$_{10kDa}$ and the other parameters namely TEP, biopolymer concentration, and chlorophyll a concentration and algal cell density during the growth phase. As illustrated in Figure 4.5, the biopolymer concentration (polysaccharides and proteins) and TEP concentration were found to correlate better with the MFI-UF$_{10kDa}$ in growth and stationary/decline phase of algae. This observation is attributed to the fact that biopolymer concentration, TEP, and MFI-UF continued to increase during the stationary/decline phase of the algae as shown in Figure 4.4. The poor correlation between algal cell density and chlorophyll-a with the MFI-UF$_{10kDa}$ during the stationary/decline phase indicate that these parameters are not fully adequate to predict/quantify the fouling potential of RO (and UF) feed water during an algal bloom (Villacorte et al., 2015c). Consequently, continuous monitoring algal cell densities and chlorophyll-a, e.g., with MODIS need to be complemented by recently developed parameters such as TEP, MFI-UF, and biopolymer concentration to get an adequate indication of the fouling properties of seawater during algal growth and blooms.

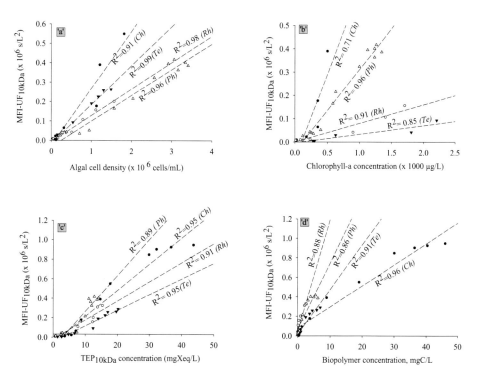

Figure 4.5: Correlation between Modified Fouling Index and algal cell density, Chlorophyll-a concentration, TEP concentration, and Biopolymer concentration for four algal species Chaetoceros affinis (Ch), Rhodomonas balthica (Rh), Tetraselmis suecica (Te), and Phaeocystis globulosa (Ph)

Furthermore, to understand the role of algal cells, the fouling potential of algal suspension without AOM was calculated using the model Equation 4.5. The measured and calculated MFI values for the four different algal species are presented in Table 4.4. The result illustrates that MFI-UF values from the experiments (algae and AOM) were much higher than theoretical calculation (algae only). This indicates that the contribution of algal cells itself without AOM to the fouling potential is very low. However, it cannot be excluded that AOM, in particular, TEP, attached to algal cells, substantially increases the specific resistance of deposited algal cells and consequently contribute to the MFI-UF$_{10kDa}$

Table 4.4: Measured and calculated MFI-UF$_{10kDa}$

Algal species	Shape	Average size, μm	Algal cells density (cells/mL)	Measured [a](MFI-UF s/L^2)	Calculated [b](MFI-UF s/L^2)
				Algae + AOM	Algae
Ch	Oval cylinder	9	1,800,000	946,000	19
Rh	Cone	10	1,600,000	324,000	22
Te	Ovoid	10	1,500,000	290,000	17
Ph	Spherical	6	3,400,000	417,000	24

(a) Calculated from experimental data (batch cultures); (b) Calculated from theoretical data: $\varepsilon = 0.4$, $\varphi = 1$, dp = average size of algal species, Cp = algal cell density, cells/mL

4.3.3 Comparison of fouling potential of AOM and AOM + algal cells

The fouling potential of AOM with and without algal cells was measured using parameters such as biopolymers, TEP$_{10\,kDa}$, MFI-UF$_{10kDa}$, MFI, and SDI (see Table 4.5). The AOM was separated from algal cells by pre-filtration of the sample through 5 μm filter. The measured values of AOM without algal cells were significantly lower for all parameters when compared to AOM with algal cells. This indicates that a part of the AOM is attached to algal cells. The AOM samples after filtration through 5 μm filter were used for the ultrafiltration experiments.

Table 4.5: Measured MFI-UF, TEP, biopolymer concentration for AOM + algal solution, and AOM solutions

Algal species	Biopolymers (mg-C/L)		TEP$_{10kDa}$ (mgXeq/L)		MFI-UF$_{10kDa}$ (s/L^2 x 1000)		MFI$_{0.45}$ (s/L^2) *		SDI3*	
	A	B	A	B	A	B	A	B	A	B
Ch	46.0	14.6	43.8	19.8	946	369	nm	110	nm	24.6
Rh	2.8	0.9	15.4	8.9	324	177	nm	80	nm	22.6
Te	7.0	4.0	20	15.8	290	272	nm	60	nm	22.5
Ph	6.0	0.4	13.4	3.2	386	21	nm	50	nm	18.6

*Samples were 50 times diluted with filtered seawater

A = AOM with algal cells, B = AOM without algal cells, nm = not measured

4.3.4 Characterization of algal organic matter (AOM)

4.3.4.1 Liquid chromatography - organic carbon detection (LC-OCD)

The organic carbon detector (OCD) chromatograms of four AOM samples obtained from LC-OCD analysis are presented in Figure 4.6a. Two distinct peaks were observed in the chromatogram. The peak that eluted approximately at 32 - 35 minutes after the sample injection into the system was assigned to the biopolymer fraction of organic carbon (high molecular weight > 20 kDa) which comprises proteins and polysaccharides. A remarkable difference was observed in the peaks of the biopolymer fraction of the four different cultured AOM solutions. In terms of absolute concentration, AOM of *Ch* has the highest biopolymer concentration (14.6 mg-C/L), which is approximately 4, 17 and 20 times higher than AOM of *Te, Rh*, and *Ph*, respectively (Figure 4.6 b).

The second peak that eluted between 45-55 minutes was assigned to building blocks and humic substances. These peaks might have originated entirely from the medium, i.e., EDTA added to the algal culture, which was used as a chelating agent to minimize the precipitation of metals in the medium. The LC-OCD analysis of EDTA showed a peak at a retention time (50 - 57 minutes) (Villacorte et al., 2015b) similar to the elution time for building blocks and humic substances (result not shown here).

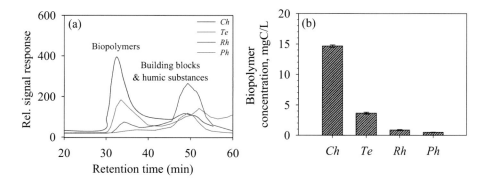

Figure 4.6: a) LC-OCD chromatograms of AOM extracted during the stationary/decline phase of the four marine algal species and b) concentration of biopolymer measured in four algal AOM

4.3.4.2 Fluorescence excitation-emission matrix (F-EEM)

The F-EEM spectra of AOM of the four algal species were performed to investigate the presence of protein-like and humic/fulvic-like substances in AOM solutions. Polysaccharides are non-fluorescent organic substances. As illustrated in Figure 4.7, both protein-like and humic-like peaks were identified in the F-EEM spectra of AOM of the four algal species. However, for all AOM, fluorescence responses mainly originated from tyrosine-like proteins (P1). For all AOM, the observed fluorescence intensity of the protein-like peak was observed to be higher than the humic-like peaks and was ranked as *Rh > Ch > Te > Ph*.

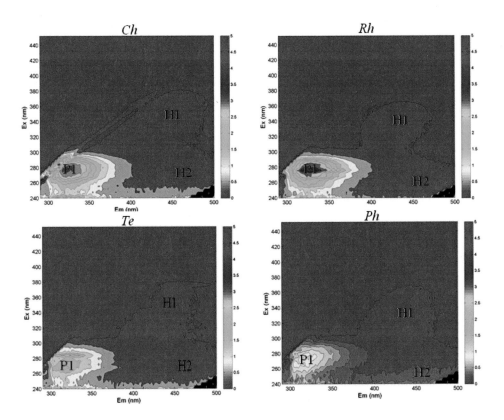

Figure 4.7: Typical F-EEM spectra prepared using Matlab for AOM samples from four algal species a) Ch. b) Rh. c) Te. and d) Ph. Legend: H1= primary humic-like peak; H2=secondary humic-like peak and P1=tyrosine-like protein peak

4.3.5 Fouling behavior of AOM generated by the four algal species in UF

Figure 8 shows the development of transmembrane pressure (TMP) versus time during the filtration cycles. From these curves, the total, backwashable and non-backwashable fouling rate has been derived (See Table 6). AOM of *Rh* and *Ch* showed a fast increase in total fouling during consecutive filtration cycles, exhibit a remarkably high backwashable fouling, which indicates that the AOM present is less sticky than present in *Te* and *Ph*. The concentrations of TEP presented in table 6 support this assumption. Figure 8 b showed that the non-backwashable fouling developed according to more or less linear in time. *Rh* showed the highest and *Ph* showed the lowest non-backwashable fouling rate.

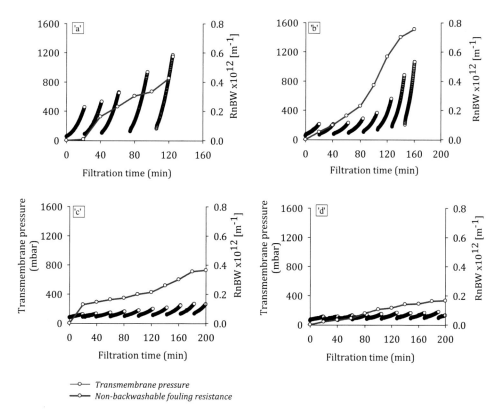

Figure 8: *Pressure and non backwashable fouling resistance development during multiple ultrafiltration cycles with AOM (C_{feed} = 0.5 mg-biopolymer-C/L) extracted from four different marine algal species (a) Chaetoceros affinis (Ch), (b) Rhodomonas balthica (Rh), (c)Tetraselmis suecica (Te), and (d) Phaeocystis globulosa*

Table 6: Fouling parameters (MFI-UF), TEP, fouling rate development, and projected chemical cleaning interval

Algal species	[a]MFI-UF$_{150kDa}$ (s/L^2 × 1000)	Total fouling rate (f$_T$), (bar/h)	Non-backwashable fouling rate, (f$_{nBW}$), (bar/h)	Backwashable fouling rate (f$_{BW}$), (bar/h)	Specific TEP (mg-Xeq/L/mg-C/L)	Projected chemical cleaning interval (h)
Rh	46	0,37	0,07	0,30	4.6	5,7
Ch	26	0,53	0,05	0,48	1,9	8,0
Te	9	0,14	0,03	0,11	2,2	13,3
Ph	1	0,03	0,01	0,02	1	40,0

[a] Calculated based on the minimum slope of the first filtration cycle.

The non-backwashable fouling rate is related to the MFI-UF$_{150kDa}$. The higher the MFI, the higher the non-backwashable fouling rate. The total fouling rate shows a similar development as well, except for *Ch*. In full-scale desalination Ultrafiltration plants, chemically enhanced backwashing (CEB) and sometimes by cleaning-in-place (CIP) is applied to restore the membrane resistance. The CEB protocol recommended by the membrane manufacturer includes cleaning with a solution (pH = 11 - 12) prepared with sodium hydroxide (NaOH) and Sodium hypochlorite (NaOCl) (200 mg/L), and with a solution (pH = 2.5 - 3.5) prepared with HCl. The projected CEB interval is based on the criterion that Chemical Enhanced Backwash (CEB) is recommended at ≥ 0.4 bar pressure increase.

4.3.6 Nature of AOM deposited on UF membranes

Figure 9 shows FTIR spectra of a virgin new and a fouled UF membrane and a fouled membrane after sonication. Moreover, It should not be excluded the difference in growth medium and conditions applied during the cultured of four algal species (see section 2.1) while comparing between four AOM samples.

The identified functional groups based on the IR spectra and the typical organic compounds associated with AOM are shown in Table 7. All AOM fouled UF membranes show the broad and intense band of stretching O-H group at a wavelength of 3,400-3,200 cm^{-1} (Peak A). This band could be due to the presence of carboxylic acids, alcoholic and phenolic compounds usually associated with polysaccharides (Mecozzi et al., 2001). The stretching refers to the change in inter-atomic distance along the bond axis and bending refers to the change in the angle between two bonds. Another peak (F) observed at a wavelength of 1,280 - 1,200 cm⁻¹ corresponds to C-O stretching, and OH deformation of COOH is an indication of the presence of polysaccharides as well. Likewise, several other observed peaks (C, D, and E) in the wavelength range from 1,650 - 1,480 cm⁻¹ are mostly associated with the presence of proteins. Peak G indicates the presence of CH aromatic compounds possibly originated from humic-like substances. Peak H shows intense absorption bands at 1,050 cm⁻¹, which corresponds to–S=O stretching of sugar ester sulphate groups.

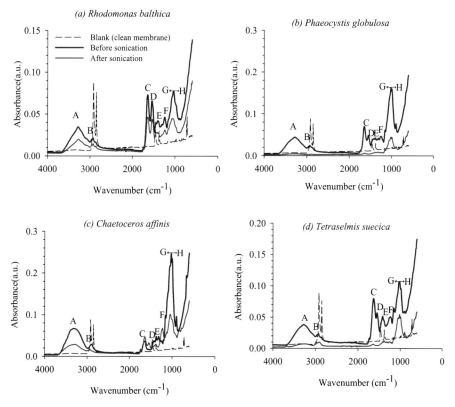

Figure 9: FTIR spectra of AOM-fouled ultrafiltration membrane samples for Rh, Ch, Te, and Ph. a) Solid line (red) represents the FTIR Spectra for AOM-fouled UF before sonication, b) Solid line (blue) represents the FTIR Spectra for AOM-fouled UF after 90 minutes of sonication at 42 kHz , and c) dotted line represents FTIR Spectra for virgin UF membrane

In general, the FTIR spectra findings were consistent with what was reported in marine mucilage aggregates by Mecozzi et al., 2001 (Mecozzi et al., 2001) and for freshwater and seawater AOM by Villlacorte et al., 2015 (Villacorte et al., 2015b). The result identifies the presence of polysaccharides, proteins, and humic substances which were also consistent with the findings of LC-OCD and F-EEM analysis.

Table 7: Absorption band, functional group, compound identified in FTIR test of AOM fouled UF

Peak	Wavelength (cm⁻¹)	Functional group	Compound
A	3,400 - 3,200[a]	Stretching OH	Polysaccharides
B	2,950 - 2,850[b]	Stretching CH_2	Lipids
C	1,650 - 1,640[a]	Stretching C=O and C-N (Amide I)	Proteins
D	1,545 - 1,540[b]	Stretching C-N & bending NH (Amide II)	Proteins
E	1,500 - 1,480[b]	C-N stretch and bend	Proteins
F	1,280 - 1,200[b]	Stretching C-O & OH deformation of COOH	Polysaccharides
G	1080 - 1070[a]	CH aromatic	Humic substances
H	1,050[a]	Stretching – S = O	Sugar ester sulphates

Interpretation of IR spectra was based on Mecozzi et al., 2001 (Mecozzi et al., 2001) and Villlacorte et al., 2015 (Villacorte et al., 2015b), a = Intense band and b = Weak band

After sonication, absorbance (peak height) was largely reduced for most peaks. Two broad and intense high peaks (A and H) linked to polysaccharides and sugar ester sulphates, respectively were just partly reduced even after extended sonication. The reduction of most of the peaks from fouled membranes after sonication (physical cleaning) was better for AOM of *Te* and *Ph* than for AOM of *Rh* and *Ch*. This indicates that the adherence of AOM to the membrane surface varied between algal species. Consequently, the non-backwashable of ultrafiltration membranes, during an algal bloom, will be governed by the characteristics of the AOM, which depends on the algal species present.

4.3.7 Passage of biopolymers through UF membranes

The rejection of biopolymer varied with the type of algal species (Figure 10a). As illustrated in Figure 10a, the percentage passage of biopolymer fraction of AOM through UF membrane when operated without coagulation ranged from 20 – 40 %. The finding indicates that biopolymers are having a size smaller than the pores of the ultrafiltration membranes. Consequently, it cannot be excluded that a part of this compound might contribute to pore blocking resulting in non-backwashable fouling.

The comparative percentage of biopolymer passage with different pre-treatment systems reported in the literature are presented in Figure 10b (Guastalli et al., 2013, Naidu et al., 2013, Salinas - Rodriguez et al., 2009, Tabatabai et al., 2014). Moreover, the source of feed water used in each case was different. The result illustrated that the existing pre-treatment systems are capable of rejecting approximately 60 - 80 % of biopolymer fraction of AOM depending upon the type of pre-treatment systems and the amount of coagulant dose applied during the operation. It cannot be excluded that the fraction of AOM passing through the pre-treatment systems causes organic fouling in SWRO membranes. It serves as a *"conditioning layer,"*an ideal attachment site for bacteria, where bacteria can grow and multiply to form biofilm in the expense of available nutrients in the SWRO feed water (Bar-Zeev et al., 2012, Berman et al., 2005, Villacorte et al., 2009). This indicates that there is a need of robust advanced pre-treatment systems that can provide acceptable feed water quality for SWRO systems.

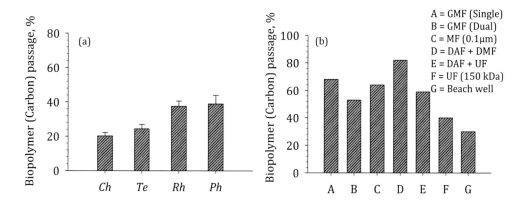

Figure 4.10: Percentage of biopolymer passage through (a) ultrafiltration membrane (150 kDa) when AOM of four cultured marine algal species used as a feed solution, and (b) various SWRO pre-treatment systems tested with different water source as feed water and varying operating conditions

4.4   Conclusions

- During the growth phase of the four different marine algal species, a linear relation was observed between MFI-UF$_{10kDa}$ and algal density, chlorophyll-, biopolymer- and TEP concentration. After the growth phase (stationary/decline phase) this relation was no longer observed.

- Substantial differences in production of biopolymers and TEP were observed between the four algal species.

- The specific membrane fouling potential measured as MFI-UF$_{10kDa}$ per mg-biopolymer-C/L for the different algal species varied by a factor 5.

- The level of the measured MFI-UF$_{10kDa}$ has to be attributed to the biopolymer/TEP and not to the algal cells itself. However, biopolymers/TEP attached to the algal cells might (substantially) contribute to the MFI-UF.

- The backwashable fouling of the ultrafiltration membranes, fed with water having 0.5 mg-biopolymer-C/L originated from the four algal species, coincide with the MFI-UF$_{150kDa}$ for *Rh, Te and Ph.* Backwashable fouling for *Ch* deviated and was substantially higher.

- The non–backwashable fouling of the ultrafiltration membranes varied strongly with the type of algal species and coincided with MFI-UF$_{150kDa}$ and TEP concentration. The higher these levels, the higher the non-backwashable fouling. *Rh* demonstrated the highest and *Ph* the lowest non-backwashable fouling (at a level of 0.5 mg-biopolymer-C/L.

- The non-backwashability is attributed to stretching -OH groups in polysaccharides and stretching -S=O groups sugar ester group present in the biopolymers.

- Biopolymers were rejected by UF (150 kDa MWCO) membranes for 60 % to 80 % depending upon on the algal species. Indicating that biopolymers having a size smaller than the pores of the ultrafiltration membranes might contribute to non-backwashable fouling.

4.5   Recommendations

- Determining MFI-UF$_{10kDa}$, biopolymer, TEP concentration, during algal blooms monitored by the MODIS satellite, will generate useful information about fouling potential of seawater at different locations.

4.6   Acknowledgements

The authors would like to thank the participants of the research theme 'Biofouling' for the fruitful discussions. The authors also appreciate Marco Dubbeldam from stitching Zeeschelp B.V. for his support in algal culture, Mieke Kersaan-Haan from Wetsus for performing the LC-OCD analysis, and Ranjita Bose from TU Delft Aerospace Engineering for performing FTIR analysis.

4.7   References

Arevalo, J., Kennedy, M.D., Salinas-Rodriguez, S.G., Sandin, R., Rogalla, F. and Monsalvo, V.M. (2017) Pretreatment systems in desalination plants to reduce extreme events impact in drinking water production., Pan-European Symposium Water and Sanitation Safety Planning and Extreme Weather Events,Bilthoven, Netherlands.

Bar-Zeev, E., Berman-Frank, I., Girshevitz, O. and Berman, T. (2012) Revised paradigm of aquatic biofilm formation facilitated by microgel transparent exopolymer particles. Proc Natl Acad Sci U S A 109(23), 9119-9124.

Berman, T. and Holenberg, M. (2005) Don't fall foul of biofilm through high TEP levels. Filtration & Separation 42(4), 30-32.

Berman, T., Mizrahi, R. and Dosoretz, C.G. (2011) Transparent exopolymer particles (TEP): A critical factor in aquatic biofilm initiation and fouling on filtration membranes. Desalination 276(1–3), 184-190.

Boerlage, Ś.F.E., Kennedy, M.D., Aniye, M.p., Abogrean, E.M., El-Hodali, D.E.Y., Tarawneh, Z.S. and Schippers, J.C. (2000) Modified Fouling Indexultrafiltration to compare pretreatment processes of reverse osmosis feedwater. Desalination 131(1), 201-214.

Bonnélye, V., Guey, L. and Del Castillo, J. (2008) UF/MF as RO pre-treatment: the real benefit. Desalination 222(1–3), 59-65.

Caron, D.A., Garneau, M.-È., Seubert, E., Howard, M.D.A., Darjany, L., Schnetzer, A., Cetinić, I., Filteau, G., Lauri, P., Jones, B. and Trussell, S. (2010) Harmful algae and their potential impacts on desalination operations off southern California. Water Research 44(2), 385-416.

Fogg, G. (1983) The ecological significance of extracellular products of phytoplankton photosynthesis. Botanica Marina 26 (1), 1-43.

Guastalli, A.R., Simon, F.X., Penru, Y., de Kerchove, A., Llorens, J. and Baig, S. (2013) Comparison of DMF and UF pre-treatments for particulate material and dissolved organic matter removal in SWRO desalination. Desalination 322, 144-150.

Henderson, R.K., Baker, A., Parsons, S.A. and Jefferson, B. (2008) Characterisation of algogenic organic matter extracted from cyanobacteria, green algae and diatoms. Water Research 42(13), 3435-3445.

http://nordicmicroalgae.org/taxon/Rhodomonas%20baltica (2017) Nordic microalgae and aquatic protozoa.

http://www.ccap.ac.uk/ccap-search.php (2017) Culture collection of algae and protozoa.

http://www.waterman.hku.hk (2016) Assessing the occurrence of an algal bloom - Chlorophyll-a concentration.

Huber, S.A., Balz, A., Abert, M. and Pronk, W. (2011) Characterisation of aquatic humic and non-humic matter with size-exclusion chromatography – organic carbon detection – organic nitrogen detection (LC-OCD-OND). Water Research 45(2), 879-885.

Kennedy, M.D., Muñoz - Tobar, F.P., Amy, G.L. and Schippers, J.C. (2009) Transparent exopolymer particles (TEP) fouling of ultrafiltration membrane systems. Desalination and Water Treatment 6 (1-3), 169 - 176.

Ladner, D.A., Vardon, D.R. and Clark, M.M. (2010) Effects of shear on microfiltration and ultrafiltration fouling by marine bloom-forming algae. Journal of Membrane Science 356(1–2), 33-43.

Leenheer, J.A. and Croué, J.-P. (2003) Peer reviewed: Characterizing aquatic dissolved organic matter. Environmental Science & Technology 37(1), 18A-26A.

Mecozzi, M., Acquistucci, R., Di Noto, V., Pietrantonio, E., Amici, M. and Cardarilli, D. (2001) Characterization of mucilage aggregates in Adriatic and the Tyrrhenian Sea: structure similarities between mucilage samples and the insoluble fractions of marine humic substance. Chemosphere 44(4), 709-720.

Merle, T., Dramas, L., Gutierrez, L., Garcia-Molina, V. and Croué, J.-P. (2016) Investigation of severe UF membrane fouling induced by three marine algal species. Water Research 93, 10-19.

Myklestad, S.M. (1995) Release of extracellular products by phytoplankton with special emphasis on polysaccharides. The Science of The Total Environment 165(1–3), 155-164.

Naidu, G., Jeong, S., Vigneswaran, S. and Rice, S.A. (2013) Microbial activity in biofilter used as a pretreatment for seawater desalination. Desalination 309, 254-260.

Pankratz, T. (2008) Red tides close desal plants. Water Desalination Report 44 (1).

Passow, U. (2002) Transparent exopolymer particles (TEP) in aquatic environments. Progress in Oceanography 55(3), 287-333.

Passow, U. and Alldredge, A.L. (1995) A dye-binding assay for the spectrophotometric measurement of transparent exopolymer particles (TEP). Limnology Oceanography 40(7), 1326-1335.

Peldszus, S., Hallé, C., Peiris, R.H., Hamouda, M., Jin, X., Legge, R.L., Budman, H., Moresoli, C. and Huck, P.M. (2011) Reversible and irreversible low-pressure membrane foulants in drinking water treatment: Identification by principal component analysis of fluorescence EEM and mitigation by biofiltration pretreatment. Water Research 45(16), 5161-5170.

Qu, F., Liang, H., Tian, J., Yu, H., Chen, Z. and Li, G. (2012) Ultrafiltration (UF) membrane fouling caused by cyanobacteria: Fouling effects of cells and extracellular organics matter (EOM). Desalination 293, 30-37.

Reddy, V. (2009) Red Tide in the Arabian Gulf. MEDRC Watermark 40(3).

Richlen, M.L., Morton, S.L., Jamali, E.A., Rajan, A. and Anderson, D.M. (2010) The catastrophic 2008–2009 red tide in the Arabian Gulf region, with observations on the identification and phylogeny of the fish-killing dinoflagellate Cochlodinium polykrikoides. Harmful Algae 9(2), 163-172.

Salinas-Rodriguez, S.G., Amy, G.L., Schippers, J.C. and Kennedy, M.D. (2015) The Modified Fouling Index Ultrafiltration constant flux for assessing particulate/colloidal fouling of RO systems. Desalination 365, 79-91.

Salinas - Rodriguez, S.G., Kennedy, M.D., Schippers, J.C. and Amy, G.L. (2009) Organic foulants in estuarine and bay sources for seawater reverse osmosis - Comparing pre-treatment processes with respect to foulant reductions. Desalination and Water Treatment 9, 155-164.

Schippers, J.C. and Verdouw, J. (1980) The modified fouling index, a method of determining the fouling characteristics of water. Desalination 32, 137-148.

Schurer, R., Janssen, A., Villacorte, L.O. and Kennedy, M.D. (2012) Performance of ultrafiltration & coagulation in an UF-RO seawater desalination demonstration plant. Desalination and Water Treatment 42(1-3), 57-64.

Schurer, R., Tabatabai, A., Villacorte, L., Schippers, J.C. and Kennedy, M.D. (2013) Three years operational experience with ultrafiltration as SWRO pre-treatment during algal bloom. Desalination and Water Treatment 51(4-6), 1034-1042.

Tabatabai, S.A.A., Schippers, J.C. and Kennedy, M.D. (2014) Effect of coagulation on fouling potential and removal of algal organic matter in ultrafiltration pretreatment to seawater reverse osmosis. Water Research 59, 283-294.

Tian, J.-y., Ernst, M., Cui, F. and Jekel, M. (2013) Correlations of relevant membrane foulants with UF membrane fouling in different waters. Water Research 47(3), 1218-1228.

Villacorte, L.O. (2014) Algal blooms and membrane-based desalination technology, Ph.D. thesis, UNESCO-IHE/TUDelft, Delft.

Villacorte, L.O., Ekowati, Y., Calix-Ponce, H.N., Schippers, J.C., Amy, G.L. and Kennedy, M.D. (2015a) Improved method for measuring transparent exopolymer particles (TEP) and their precursors in fresh and saline water. Water Research 70, 300-312.

Villacorte, L.O., Ekowati, Y., Neu, T.R., Kleijn, J.M., Winters, H., Amy, G., Schippers, J.C. and Kennedy, M.D. (2015b) Characterisation of algal organic matter produced by bloom-forming marine and freshwater algae. Water Research 73, 216-230.

Villacorte, L.O., Ekowati, Y., Winters, H., Amy, G., Schippers, J.C. and Kennedy, M.D. (2015c) MF/UF rejection and fouling potential of algal organic matter from bloom-forming marine and freshwater algae. Desalination 367, 1-10.

Villacorte, L.O., Ekowati, Y., Winters, H., Amy, G., Schippers, J.C. and Kennedy, M.D. (2013) Characterisation of transparent exopolymer particles (TEP) produced during algal bloom: a membrane treatment perspective. Desalination and Water Treatment 51((4-6)), 1021-1033.

Villacorte, L.O., Kennedy, M.D., Amy, G.L. and Schippers, J.C. (2009) The fate of Transparent Exopolymer Particles (TEP) in integrated membrane systems: Removal through pre-treatment processes and deposition on reverse osmosis membranes. Water Research 43(20), 5039-5052.

Villacorte, L.O., Tabatabai, S.A.A., Dhakal, N., Amy, G., Schippers, J.C. and Kennedy, M.D. (2015d) Algal blooms: an emerging threat to seawater reverse osmosis desalination. Desalination and Water Treatment 55(10), 2601-2611.

Voutchkov, N. (2010) Considerations for selection of seawater filtration pretreatment system. Desalination 261(3), 354-364.

5

The role of tight ultrafiltration on reducing fouling potential of SWRO feed water

Contents

This chapter is based on:

Dhakal, N., Salinas Rodriguez, S.G., Schippers, J.C. and Kennedy, M.D. (2017), Role of tight ultrafiltration on reducing fouling potential of SWRO feed water. *In preparation for Desalination*

Abstract

The failure of conventional pre-treatment to provide acceptable feed water quality for seawater reverse osmosis (SWRO) during algal blooms underlines the significance of robust pre-treatment systems. This study investigated the effectiveness of tight ultrafiltration (10 kDa) membrane as pre-treatment in delaying the onset of organic/biological fouling in SWRO feed water during algal blooms. The proof of principle experiments were performed in laboratory and pilot plant using various MF/UF membranes and algal organic matter (AOM) produced by Chaetoceros affinis as a feed solution. The feed and permeate of MF/UF membranes were analyzed in terms of biopolymer concentration, and bacterial regrowth potential (BRP). Furthermore, biofouling experiments were also performed using membrane fouling simulators (MFS) to simulate biofouling in spacer-filled RO membrane channels.

Results illustrated that the rejection of algal biopolymer produced by Chaetocerous affinis was 3 to 4 times higher with tight UF (10 kDa) compared to the high molecular weight cut off MF/UF membranes. The lower biopolymer concentration in permeate coincide with the lower bacterial regrowth potential. The relationship between the bacterial regrowth and biopolymer concentration was found linear with $R^2 = 0.88$. Moreover, no substantial difference was observed in measured net bacterial regrowth in permeate collected from tight UF (10 kDa) and standard UF (150 kDa) from pilot experiments. It could be attributed to the contribution of passage of low molecular weight organics from both UF membranes. The biofouling experiments performed using MFS monitor fed with permeate of 150 kDa and 10 kDa UF at a cross flow velocity of 0.2 m/s also showed no substantial increase in the feed channel pressure drop in the MFS monitor. Moreover, the result of membrane autopsy showed biomass accumulation of 860 pg ATP/cm^2 in MFS fed with 10 kDa UF permeate, which was 2 times lower than in MFS fed with 150 kDa UF permeate. Overall, the results illustrated the potential of tight UF membranes towards delaying the occurrence of biofouling in SWRO membranes. Nevertheless, the non-backwashable fouling rates development after each succeeding CEB cycles were approximately 1.5 times higher for 10 kDa UF compared to 150 kDa UF. Therefore, it is still important to improve the backwashability performance of the tight UF membrane for the better future application. It is expected that improving the surface porosity of the membrane can better remove the cake/gel layer formed on the membrane surface during backwashing/CEB and improves the backwashability of the membranes.

Keywords: *Membrane fouling, tight ultrafiltration, conventional ultrafiltration, algal blooms, algal organic matter*

5.1   Introduction

Algal blooms are increasingly being reported as one of the causes of operational problems in seawater reverse osmosis (SWRO) desalination plants (Caron et al., 2010, Petry et al., 2007, Villacorte et al., 2015c). During algal blooms, algae release algal organic matter (AOM) into the water treatment process (Babel et al., 2010).The larger fraction of AOM consists of biopolymer (polysaccharides and proteins) (Myklestad et al., 1972), which contains a sticky and gel-like material called transparent exopolymer particles (TEP) (Mopper et al., 1995). This fraction of AOM has been identified as the most problematic foulants (Zhang et al., 2015), causing organic fouling in SWRO membranes. The accumulated TEPs in the membrane surface can serve as a *"conditioning layer"*, an ideal attachment site for bacteria, where bacteria can grow and multiply to form biofilm in the expense of available nutrients in the SWRO feed water (Bar-Zeev et al., 2012, Berman et al., 2005, Villacorte et al., 2009).When the accumulation of biofilm reaches a certain threshold, biofouling problems are expected in the membrane systems (Flemming, 2002) The consequences of biofouling are i) decreased membrane permeability, ii) increased pressure drop along the spacer channel, iii) increased frequency of chemical cleaning, and iv) possible increase in the replacement frequency of membranes (Matin et al., 2011, Radu et al., 2012) or eventually shut down the operation of SWRO (Pankratz, 2008, Reddy, 2009).

Achieving stable SWRO feed water during algal blooms and prevent the occurrence of organic/biofouling is a challenge in seawater desalination membranes systems. Flemming (2011) suggested the most promising strategy for biofouling control is to minimize the bacterial concentration and limit the nutrient concentration (C, N, and P) in RO feed water using a reliable pretreatment system (Flemming, 2011). The selection of pre-treatment is a site-specific which depends on the feed water quality, and membrane types (Abd El-Aleem, 1998). In practice, pre-treatment can be conventional (coagulation/flocculation, sedimentation or flotation and granular media filtration (single/dual) or advanced using microfiltration/ultrafiltration (MF/UF). Moreover, several studies suggested that the existing pre-treatment systems are effective in removing the algae but they still allow the passage of a substantial fraction (40 – 80 %) of algal organic matter (AOM) measured in terms of biopolymer (Guastalli et al., 2013, Naidu et al., 2013, Salinas - Rodriguez et al., 2009, Tabatabai et al., 2014). The passage of algal biopolymer depends upon the type of pre-treatment used, and a dose of coagulant applied. This highlights that the current pre-treatment systems are not adequate to prevent membrane fouling especially biofouling in SWRO membranes.

The failure of conventional pre-treatments to provide acceptable feed water quality for SWRO during algal blooms has gained the attention of desalination industry to look for alternative advanced pre-treatment systems (Villacorte et al., 2015b). It is expected that the tight ultrafiltration membrane with molecular weight cut-off (MWCO) of 10 kDa can remove the significant fraction of AOM from SWRO feed water and thus supports in delaying the occurrence of biofouling in SWRO systems. Therefore, the main goal of this study was to compare the effectiveness of tight ultrafiltration (10 kDa) membrane and standard MF/UF membranes (flat sheets: 0.1µm, 100 kDa, 30 kDa and hollow fiber: 150 kDa) in reducing the biofouling potential of SWRO feed water during algal blooms. Algal bloom conditions were achieved by culturing a marine bloom-forming algae "*Chaetoceros affinis*" in both laboratory and pilot studies. The specific objectives were;

i.   To perform the proof of principle tests in laboratory scale by evaluating the permeate quality of tight ultrafiltration (10 kDa) and high molecular weight cut off MF/UF (flat membranes: 0.1 µm, 100 kDa, 30 kDa and hollow fiber: 150 kDa) membranes in terms of;

-Biopolymer concentration (measured by LC-OCD),

-Bacterial regrowth potential (using flow cytometry),

-Modified fouling index (MFI-UF$_{10kDa}$), and transparent exopolymer particles (TEP$_{10kDa}$).

ii. To investigate in a pilot scale experiments the effectiveness of hollow fiber tight UF (10 kDa) and standard UF (150 kDa) membranes towards delaying the onset of organic/biological fouling in SWRO feed water during algal blooms.

-Feed and permeate quality regarding AOM rejection and bacterial regrowth potential (BRP) reduction

-Feed channel pressure drop (ΔP) development and biomass accumulation in membrane fouling simulator (MFS) when fed with permeate of permeate of two different UF membranes

-Hydraulic performance of UF membranes in terms of backwashability

5.2   Material and methods

5.2.1 Algal culture, AOM extraction, and characterization

A common bloom-forming marine diatom *Chaetoceros affinis* was grown to represent the algal bloom situation in seawater. The algal strain (CCAP 1010/27) was purchased from the Culture Collection of Algae and Protozoa (SAMS, Scotland). The algal strain was inoculated in sterilized artificial seawater (ASW) spiked with nutrients and trace elements based on Guillard's (f/2 + Si) medium. The ASW that represents the ionic composition of the North Seawater (TDS 34 g/L, pH 8 ± 0.2) was prepared using the typical inorganic salts. The composition of prepared Guillard's (f/2+Si) medium and ASW is presented in the supplementary data S1. The cultures were incubated at 20°C under controlled light (12 h light: 12 h dark) condition using mercury fluorescent lamps with an average incident photon flux density of 40 $\mu mol.m^{-2}.s^{-1}$. The cultures were continuously mixed on a shaker for approximately two weeks. The algal cultures flasks were removed from the shaker and allowed to settle for 24 hours. The supernatant from the culture flask was extracted and filtered using 5 μm polycarbonate filter (Whatman Nuclepore) to remove the remaining algal cells in the suspension. The extracted algal organic matter (AOM) solution was stored at 5 °C and sent to DOC-Labor (Karlsruhe, Germany) and Wetsus, Leeuwarden for LC-OCD characterization. The extracted AOM solutions were used for the proof of principle experiments in the laboratory scale.

The algal production was scaled up for the demonstration experiments in a pilot plant. For this purpose, *Chaetoceros affinis* was cultured in 250 L plastic bags using 2 μm filtered, and UV disinfected natural North Seawater (Figure 5.1). In short, the raw North Seawater after 2 μm glass filter and UV disinfection was pumped to each plastic bags at a flow rate of 1 L/h. The nutrient (f/2+Si) was periodically supplied inside the plastic bag. Air was continuously supplied for mixing the added nutrients homogeneously. Furthermore, the bag was also supplied with CO_2 to adjust the pH to a level of ~ 8.5. As illustrated in Figure 5.1, there were 6 bags in parallel with the capacity of 6 - 7 L/h of AOM production. The AOM was continuously harvested from the supernatant of the bags and collected in a tank. The LC-OCD analysis of the collected

AOM and raw North seawater was performed at Wetsus, Leeuwarden. Two solutions (AOM and raw North seawater) were hydraulically mixed to have a final AOM concentration of 0.5 mg - biopolymer - C/L in ultrafiltration feed tank.

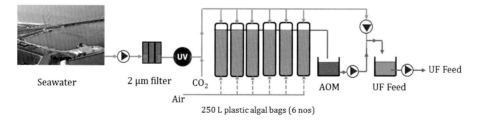

Figure 5.1: Algal culture, AOM production from marine diatom Chaetoceros affinis in pilot study

5.2.2 Proof of principle in laboratory

An overview of the experimental approach is presented in Figure 5.2, which shows the AOM rejection experiments performed in two phases; i) with flat-sheet MF/UF membranes and ii) with hollow fiber UF membranes. The feed solutions considered in both phases of experiments were AOM produced by *Chaetoceros affinis* pre-filtered with 0.22 µm Polyvinylidene difluoride (PVDF) and diluted with autoclaved ASW (121 °C, 20 minutes). The feed and permeate samples after each AOM rejection experiments were analyzed in terms of biopolymer concentration (using LC-OCD analysis) and bacterial regrowth potential (using flow cytometry).

AOM rejection by flat sheet MF/UF membranes

MF/UF flat membranes with different pore sizes, namely: 0.1 µm polycarbonate (PC, Whatman), 100 kDa, 30 kDa, and 10 kDa regenerated cellulose (RC, Millipore), were used for AOM rejection experiments. Before the filtration experiments, all membranes were soaked in ASW for 24 hours and flushed with MilliQ water to remove possible organic contaminants that might release by a new virgin membrane. Filtration through 0.1 µm PC membrane was performed in a vacuum filtration systems with < 0.2 bars driving vacuum pressure, while filtration through RC membranes was carried out at constant flux (60 L/m²/h) using a syringe pump (Harvard pump 33). The first 20 mL of the filtered volume was discarded, and then a total volume of 125 mL permeate samples were collected in a clean glass bottle for analyses.

AOM rejection by hollow fiber UF membranes

Hollow-fiber ultrafiltration polyethersulfone (PES) membrane with molecular weight cut-off 150 kDa and 10 kDa were used for AOM rejection experiments. The membrane modules were prepared in a laboratory using six numbers of capillary ultrafiltration membrane fibers (Pentair X-Flow, the Netherlands) with an internal diameter of 0.8 mm and an effective filtration length of 30 cm. The membrane pen module preparation was according to the protocol described elsewhere (Tabatabai, 2014). In short, the membrane fibers were potted inside 8 mm (outer diameter) polyethylene tubing (Festo, Germany) using polyurethane glue (Bison, The Netherlands). The effective membrane surface area was 45 cm^2 ± 2 %. The prepared membrane pen modules were soaked in hot water at 40 °C for 24 hours. All the entrapped air inside and between the fibers were released before the filtration experiments. The clean water membrane resistance was measured using MilliQ water. Finally, the multi-filtration cycles were performed using AOM as a feed solution using the following experimental conditions;

Filtration flux	50 L/m^2/h
Filtration mode:	inside outside and dead-end
Filtration time	30 minutes
Backwash (with permeate)	60 seconds
Backwash flux	125 L/m^2/h

The filtrate of the first filtration cycle was discarded, and then approximately 500 mL of permeate samples were collected in a clean glass bottle for analyses.

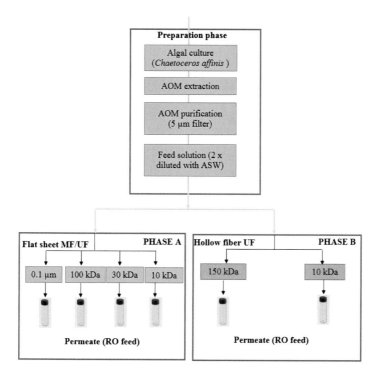

Figure 5.2: An overview of the experimental approach

5.2.3 Investigation in a pilot plant

The effectiveness of hollow fiber polyethersulphone (PES) ultrafiltration membranes (molecular weight cut-off, 150 kDa, and 10 kDa) towards biofouling reduction potential of SWRO feed water during algal blooms were investigated in a pilot testing. For this purpose, the pilot plant of Pentair X-Flow, which was located in the estuarine region of South-Western Netherlands, Zeeland Province was used. The pilot comprises of submerged open seawater intake (4 m depth), 50 μm Amiad strainer, a bench scale ultrafiltration systems as presented in Figure 5.3. Furthermore, the pilot plant also has a membrane-fouling simulator (MFS) for the biofouling experiments. The principle of ultrafiltration membrane operation and experimental conditions are described in section 5.2.3.1. Likewise, the details of MFS operation are described in section 5.2.3.2.

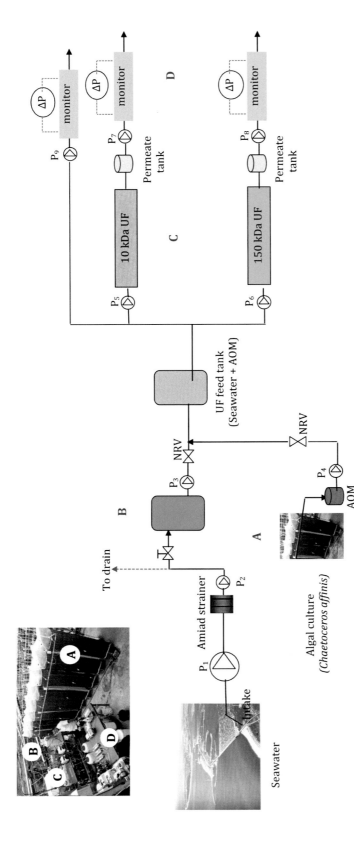

Figure 5.3: General scheme of pilot plant used in this study. Abbreviations: P_1: Intake pump; P_2: Booster pump; P_3: Masterflex pump (Seawater dosing pump); P_4: AOM dosing pump; P5 and P6 UF feed pump; P7, P8, and P9: Masterflex pump for Membrane Fouling Simulator (MFS); NRV: Non-return valve

Picture showing a general outlook and the components of pilot plants (A) algal culture of marine diatom "Chaetoceros affinis", (B) raw seawater collection tank collected after 50 μm Amiad strainer, (C) A bench scale (1") ultrafiltration unit for 150 kDa and 10 kDa UF, (D) Membrane Fouling Simulator for biofouling experiments

As illustrated in Figure 5.3, the intake structure consists of a submerged set of pipes, intake pump (P_1), and coarse screening. The intake was about 4 m in depth below the sea level. The pumped raw seawater passes through 50 μm Amiad strainer. The booster pump P_2 was used to maintain the pressure of 1.5 - 2 bars. The raw seawater (40 L/h) after Amiad strainer was collected in a tank B as shown in Figure 5.3. The remaining volume of seawater was recirculated back to the sea. In addition, a common species of blooms forming marine algae, *Chaetoceros affinis*, was cultured in 250 L plastic bags (A) as shown in Figure 5.3. This production was to create an artificial algal bloom conditions. The detail of algal culture and production is explained in section 5.2.1.

The raw seawater was pumped (with P_3) from the collected tank B and was mixed hydraulically with AOM pumped (with P_4) from an AOM collection tank. The mixing ratios of two solutions were adjusted to have the final AOM concentration of 0.5 mg-biopolymer-C/L in the UF feed tank. Two hollow fiber ultrafiltration membranes (150 kDa and 10 kDa) supplied by Pentair X-Flow (Xiga Seaguard) were operated in parallel using AOM (0.5 mg-biopolymer-C/L) as a feed solution. The details of operational parameters are presented in section 5.2.3.1 The effectiveness of two hollow fiber membranes regarding biofouling reduction was tested using membrane fouling simulator (MFS).Three MFS cells (20 cm x 1 cm) were installed after UF, one was fed directly with UF feed, and other two were fed with permeate of 150 kDa and 10 kDa UF. The RO membrane used in MFS experiments were purchased from Toray (TMC 810 C). The details are explained in Section 5.2.3.2.

5.2.3.1   Ultrafiltration experiments

Hollow-fiber polyethersulphone (PES) UF membranes (molecular weight cut-off, 150 kDa, and 10 kDa) obtained from Pentair X-Flow were used for this experiment. Membrane modules were fabricated in Pentair X-Flow, Enschede, the Netherlands by potting 120 numbers of capillary fibers (internal diameter of 0.8 mm and 95 cm effective length) inside 1inch outer diameter polyethylene tubing. The effective membrane surface area was approximately 0.28 m². The parallel experiments were performed using 150 kDa and 10 kDa UF as shown in Figure 5.3. The AOM (0.5 mg-biopolymer-C/L) was used as a feed solution during the filtration experiments. Prior to the filtration experiments, all the entrapped air inside and outside the fibers was released. The membrane was also flushed with tap water to remove any preservative materials and any release materials from the membrane itself. The clean water membrane resistance (R_m) at the filtration flux (40 L/m²/h) was measured using Tap water as a feed solution. The detail operational conditions applied for the both UF filtration experiments are summarized in Table 5.1.

Table 5.1: Operational conditions for ultrafiltration experiments

Parameters	Value
Feed concentration	0.5 mg-biopolymer-C/L
Filtration flux	40 L/m²/h
Filtration time	20 minutes
Backwash time	1 minute
Backwash flux	250 L/m²/h
CEB start	150 L/m²/h (150 kDa) & 70 L/m²/h (10 kDa)
CEB flux	125 L/m²/h
CEB dosing time	2 minutes
Soaking time	10 minutes
Coagulation	No

The principle how UF was operated in this study is presented in Figure 5.4.

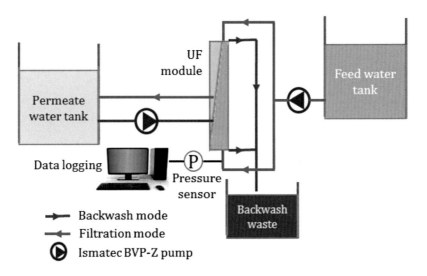

Figure 5.4: Scheme for ultrafiltration experiments

The feed water was fed from both ends, while permeate was drawn from one end of the module. The filtration was performed in inside-out mode. The permeate water of each molecular weight cut-off membrane was used for the backwashing of the respective membrane. The transmembrane pressure (TMP) development in each filtration cycle was recorded using a pressure sensor. The backwashable and non-backwashable fouling rates were calculated by plotting transmembrane pressure development over filtration time. All the experiments were performed with no coagulation.

The chemical enhanced backwash (CEB) of the ultrafiltration membrane was performed using chemicals prepared onsite using tap water. The criterion for performing CEB was the decline of permeability of the membrane below 150 L/m²/h (for 150 kDa UF) & 70 L/m²/h (for 10 kDa UF). The CEB was first performed with acid dosing (pH = 2-3), rinsed with UF permeate and soaked for 10 minutes. Secondly, the membranes were cleaned with a solution mixed with NaOH and NaOCl solution (pH = 11, chlorine concentration = 200 mg/L), rinsed again with UF permeate and soaked for 10 minutes.

5.2.3.2   Membrane fouling simulator (MFS)

Biofouling experiments were performed with membrane fouling simulators (MFS) to simulate biofouling in spacer-filled RO membrane channels. The dimensions of MFS used were 20 cm ×1 cm × 0.078 cm. The design specifications of MFS cells are according to described elsewhere (Vrouwenvelder et al., 2006). The reproducibility of MFS cells were performed and compared with theoretical calculation (see supplementary data S2).

5.2.3.2.1   Feed channel pressure drop (ΔP) development in MFS cells

A virgin RO membrane and spacer purchased from Toray (TMD 810) was used in the MFS experiments. The membrane and spacer were cut into sheets (20 cm × 1 cm) using cutter at Wetsus, Leeuwarden. The cut membrane was stored in 1 % sodium bisulfate solution for at least 24 hours and rinsed with water before the experiment. All the experiments were performed with MFS cells (no permeation) at a cross flow velocity of 0.2 m/s, which is equivalent to a flow rate of 4.8 L/h. A multi-channel peristaltic master flex pump was used to fed the feed water to MFS monitors. The pressure drop transmitter (PMD70-2071/0, Endress + Hauser) was installed to measure the feed channel pressure drop in MFS cells. The feed channel pressure drop is measured as the difference between the pressure at the feed (inlet) and (Outlet) of the MFS monitors. All data were recorded every 30 seconds and was saved in Ecograph T RSG35 data manager (Endress + Hauser) for further analysis.

5.2.3.2.2 Membrane autopsy from MFSs

Autopsies of MFSs monitors were performed at the end of the experiment. The fouled membranes, as well as the spacer, were cut into pieces with known dimensions. The sample of membrane and spacer were placed into a 50 mL tube filled with 0.22 μm filtered and autoclaved artificial seawater (40 mL). The samples were tightly covered, vortexed (Heidolph REAX 2000) for 10 seconds and sonicated (Branson 2510E-MT) for 5 minutes at a frequency of 42 kHz. The Adenosine triphosphate (ATP) and total organic carbon (TOC) were analyzed after sonication of the samples.

5.2.4 Feed and permeate quality characterization

5.2.4.1 Liquid-chromatography organic carbon detection (LC-OCD)

The quantification and fractionation of AOM were performed using the liquid chromatography organic carbon detection (LC-OCD) in DOC-Labor (Karlsruhe, Germany) and Wetsus facilities (Leeuwarden, the Netherlands). The general protocol for this technique is described by (Huber et al., 2011). Before the analyses, all samples were pre-filtered through 0.45 μm Millipore filter.

5.2.4.2 Bacterial regrowth potential (BRP)

The bacterial regrowth potential of feed and permeate samples were measured using the method described at chapter 3 of this thesis. In short, the method involves the removal/inactivation of bacteria (live + dead) from the water sample, followed by re-inoculation with live natural bacterial consortium (10^4 cells/mL), incubation at 30 °C, and monitoring of microbial growth using flow cytometry. The net live bacterial regrowth was calculated from the bacterial growth curve for the data interpretation.

5.2.4.3 Transparent exopolymer particles (TEP$_{10kDa}$)

The transparent exopolymer particles (TEP$_{10kDa}$) were measured according to the protocol described by (Villacorte et al., 2015a). The water sample was filtered through 10 kDa regenerated cellulose Millipore membrane filters. The retained TEP concentration on the filter paper was measured by staining with Alcian blue and spectrophotometric techniques and expressed in mg X_{eq}/L.

5.2.4.4   Modified fouling index (MFI-UF$_{10kDa}$)

The modified fouling index (MFI-UF$_{10kDa}$) was measured using the constant flux MFI-UF protocol developed and modified by (Boerlage et al., 2000, Salinas-Rodriguez et al., 2015). The MFI-UF was measured at a flux of 60 L/m^2/h.

5.3   Results and discussion

5.3.1 Characterization of the harvested AOM solution

Figure 5.5 presents the LC-OCD chromatograms of AOM solution (diluted and non-diluted) harvested during the stationary/death phase of cultured *Chaetoceros affinis.* The LC-OCD analyses fractionate the components of the organic material into high-molecular-weight (> 20 kDa) i.e., biopolymers (polysaccharides and proteins), medium-molecular weight such as building blocks (350 - 1000 Da) and low molecular weight component as LMW acids, and neutrals (< 350 kDa). As illustrated in Figure 5.5, the first peak eluted at a retention time of about 45 minutes was biopolymers, and other peaks eluted at a retention time higher than 85 minutes was assigned to the low molecular weight. The peaks associated with the low MW acids and neutrals might have originated from EDTA (290 Da), vitamin B12 (1355 Da), and biotin (244 Da) added to the algal culture. The addition of (Vitamin B12) might have eluted the peak that belongs to the category of LMW acids and humic substances, while the peak eluted by EDTA and Biotin belongs to the category of LMW neutrals. An LC-OCD analysis of EDTA showed a peak at a similar elution time as the LMW neutrals peak (Villacorte et al., 2015b) (results not shown here).

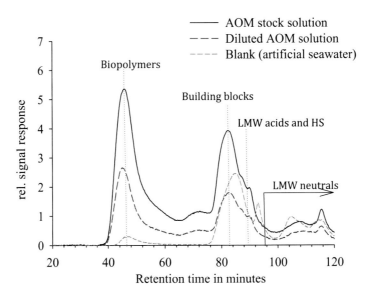

Figure 5.5: LC-OCD chromatograms of AOM diluted with ASW and non-diluted AOM solution extracted from marine bloom-forming algae species Chaetoceros affinis, and blank (artificial seawater)

Table 5.2 presents the concentration of different fractions of organic carbon in diluted and non-diluted AOM solution. Almost 40 – 50 % of the total organic carbon was a biopolymer fraction. The result was consistent with the finding of (Tabatabai et al., 2014). The quantification of AOM showed the concentration of 3.9 mg-biopolymer-C/L (non-diluted AOM) and 0.97 mg-biopolymer-C/L (diluted AOM). The diluted AOM solution was used as a feed solution for proof of principle tests perforemd in laboratory (see section 5.3.2).

Table 5.2: LC-OCD analysis of AOM stock and diluted AOM sample harvested from Chaetoceros affinis

LC-OCD organic carbon fraction	Molecular weight	Non-diluted AOM solution		Diluted AOM sample	
		Concentration	Percentage	Concentration	Percentage
	[Da]	[mgC/L]	[%]	[mgC/L]	[%]
Biopolymers	> 20,000	3.9	48.6	0.97	38.1
Humic substances	1,000-20,000	n.q.	n.q.	n.q	n.q.
Building blocks	300-500	2.4	29.3	0..55	22.9
LMW neutrals	< 350	1.7	20.7	0.85	35.4
LMW Acids	< 350	0.1	1.4	0.03	1.25

n.q. = not quantifiable (<1ppb), LMW = low molecular weight

5.3.2 Proof of principle in laboratory scale experiments

5.3.2.1   AOM rejection by flat sheet MF/UF and hollow fiber UF membranes

The MF/UF membranes are not capable of rejecting the low molecular weight fraction of AOM, thus measuring the rejection of biopolymer fraction is the better approach to understanding the rejection of AOM. As illustrated in Figure 5.6a, rejections of biopolymer fraction of AOM varied extensively with membrane pore sizes. The LC-OCD characterization showed that the relative signal response for biopolymer was decreased as the membrane pore size also decreases (Figure 5.6a). Results showed that the biopolymers fractions of AOM were largely retained by the lower pore size of the membrane (10 kDa). However, no apparent removal of low MW organics was observed irrespective of the membrane pore size. Quantitative analysis showed that the average biopolymer passage through the membranes were 95 %, 58 %, 39 %, and 20 %, respectively for 0.1 µm PVDF, 100 kDa, 30 kDa, and 10 kDa RC membranes (Figure 5.6b). Likewise, the biopolymer passage through 150 kDa and 10 kDa hollow fiber UF membranes were approximately 60 % and 10 %, respectively. Although biopolymers are reported to be larger than 20 kDa (Huber et al., 2011), we still observed 10 – 20 % passage of biopolymer through 10 kDa RC membranes, which could be due to the larger pore size variation of the RC membranes (Kim et al., 1994). Furthermore, Passow (2000) reported the algal derived biopolymers of < 3 kDa, which might have passed through 10 kDa membranes (Uta, 2000). Salinas et al, (2015) also reported that the passage of biopolymers through 100 kDa, 30 kDa, and 10 kDa membranes, when fed with natural waters, were respectively 70 %, 31 %, and 31 % (Salinas-Rodriguez et al., 2015). Moreover, Villacorte et al, (2015) reported higher biopolymer rejection by 0.1 µm, 100 kDa, and 10 kDa MWCO membranes as compared to this study (Villacorte et al., 2015b). It could be due to the higher biopolymer concentration of feed solution (~ 6 mg-C/L) used by Villacorte et al, (2015). It is expected that the feed solution with a higher biopolymer concentration cause more rapid fouling on MF/UF membranes, which also means the fast formation of a secondary rejection gel layer on the surface of the membrane, leading to a higher biopolymer rejection than on a clean membrane.

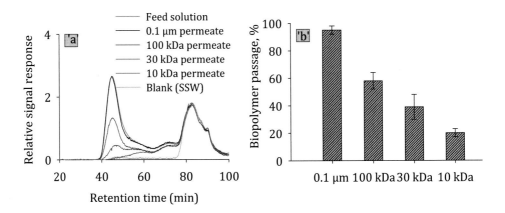

Figure 5.6: (a) LC-OCD chromatogram (measured at DOC-Labor, Germany) for feed and permeate samples after passing AOM through various pore-size MF/UF membranes and (b) biopolymer passage through various pore-size MF/UF membranes

5.3.2.2 Bacterial regrowth potential reduction by MF/UF membranes

The bacterial regrowth potential of feed and permeate samples of various pore size of flat sheet MF/UF, and hollow fiber UF membranes were investigated. The feed and permeate samples were first pre-filtered through 0.22 µm PVDF filters, re-inoculate with a live natural consortium of marine bacteria (10^4 cells/mL), incubated at 30 °C, and enumerated the bacterial regrowth using flow cytometry. The average initial bacterial cell concentration after bacterial inoculation was approximately (1 - 1.5) ×10^4 cells/mL. The growth of inoculated bacteria monitored for 27 days showed the lag phase of 24 hours followed by a rapid exponential phase and a stationary/death phase. Further, to have a better comparison, the net bacterial regrowth was calculated by subtracting the initial and the maximum bacterial cell concentration from the growth curve and plotted as shown in Figure 5.7. As illustrated in Figure 5.7a, the net bacterial regrowth reductions were 32 %, 45 %, 72 %, and 79 % for 0.1 µm, 100 kDa, 30 kDa, and 10 kDa membranes, respectively. The similar observations were also found with hollow fiber ultrafiltration membranes, which showed net bacterial regrowth reductions of 46 % and 79 %, respectively for 150 kDa and 10 kDa UF membranes (Figure 5.7b).

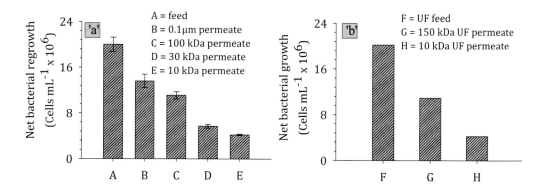

Figure 5.7: Net bacterial regrowth measured in the feed and permeate samples of a) flat sheet MF/UF membranes (0.1 μm, 100 kDa, 30 kDa, and 10 kDa), and ii) hollow fiber UF membranes 150 kDa and 10 kDa

The summary of comparative results showing the percentage of biopolymer passage and the reduction factor by various pore size MF/UF membranes in terms of biopolymer and net bacterial regrowth are presented in Table 5.3.

Table 5.3: Summary of comparative results showing the percentage of biopolymer passage and the reduction factor by various pore size MF/UF membranes

Membrane configuration	MWCO membranes	Biopolymer concentration in permeate,		NBR (10^6 cells/mL)	Reduction factor	
		mgC/L	% passage,		Biopolymer	NBR
Flat sheet MF/UF	0.1 μm PVDF	0.92 ± 0.03	95 ± 3	13.6	1.05	1.5
	100 kDa RC	0.56 ± 0.05	58 ± 6	11.1	1.73	1.8
	30 kDa RC	0.38 ± 0.08	39 ± 9	5.7	2.55	3.5
	10 kDa RC	0.19 ± 0.03	20 ± 3	4.2	5.10	4.7
Hollow fiber UF	150 kDa PES	0.70 ± 0.07	72 ± 2	11.7	1.38	1.7
	10 kDa PES	0.12 ± 0.04	12 ± 1	5	8.9	4.0

MWCO = Molecular weight cut-off; NBR =net bacterial regrowth

The measured biopolymer concentration and the net bacterial regrowth in permeate of MF/UF membranes were consolidated, and a regression analysis was performed as shown in Figure 5.8a. The result showed a linear relationship with the degree of coefficient R^2 = 0.88 between biopolymer concentration and the net bacterial regrowth. This suggested that the biopolymer indeed enhances the growth of bacteria. To investigate if biopolymer was possibly degraded or consumed by bacteria, the LC-OCD analysis was performed for feed and permeate of MF/UF flat sheet membranes collected at day 0 and day 27 of the incubation period. As illustrated in Figure 5.8b, the percentage reduction of biopolymer fraction of AOM was approximately 84 % in feed and permeates samples of 0.1 µm, 100 kDa, and 30 kDa after 27 days of incubation. Moreover, the reduction was about 50 % in the permeate sample of 10 kDa UF. This suggested that bacteria are degrading the larger biopolymers more easily than, the smaller ones. It was also reported by Passow (2002) that the biopolymer fraction in natural water could be degraded by bacteria in a matter of few hours to several months (Passow, 2002).

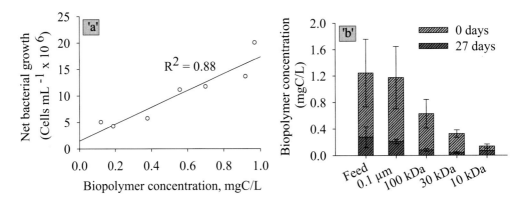

Figure 5.8: (a) Correlation between the biopolymer concentration and net bacterial regrowth (b) Biopolymer concentration measured by LC-OCD at day 0 and day 27 of incubation in the feed and permeate sample of flat sheet MF/UF membranes

5.3.2.3 TEP$_{10kDa}$ and MFI-UF$_{10kDa}$

The percentage reduction in TEP$_{10kDa}$ and MFI-UF$_{10kDa}$ by hollow fiber ultrafiltration membranes (10 kDa and 150 kDa) when fed with AOM solutions prepared with different concentration, i.e., 100 %, 50 %, 25 %, and 12.5 % were determined as shown in Figure 5.9. Various concentrations of AOM were prepared by diluting the AOM with artificial seawater (TDS = 34 g/L, pH = 8.0). As illustrated in Figure 5.9, the percentage reduction of TEP$_{10kDa}$ and MFI-UF$_{10kDa}$ measured in permeate of 150 kDa UF was approximately 10 – 20 % lower compared to the measured in permeate of 10 kDa UF. The percentage reduction in TEP$_{10kDa}$ and MFI-UF$_{10kDa}$ was found higher when a higher concentration of AOM was fed through UF membranes. Moreover, the observed non-uniform reduction of MFI-UF$_{10kDa}$ and TEP$_{10kDa}$ might be due to the variation (11.7 %) in the clean water membrane resistance (R_m) of 10 kDa RC membranes used during the measurement. It is expected that the membrane with lower membrane porosity can have higher retention of AOM leading to higher MFI-UF$_{10kDa}$ and TEP$_{10kDa}$ results, respectively.

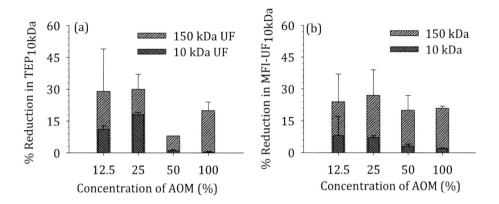

Figure 5.9: Percentage reduction of a) TEP$_{10kDa}$, b) MFI-UF$_{10kDa}$, when AOM of different concentration was fed through 10 kDa and 150 kDa UF hollow fiber membranes

5.3.3 Investigation on a pilot scale

The biofouling reduction potential of SWRO feed water during algal blooms using two hollow fiber UF membranes (150 kDa and 10 kDa) as pre-treatment were investigated in a pilot testing. The pilot plant abstract the raw seawater from the North Seawater located in Jacobahaven, the Netherlands.

5.3.3.1   Raw water quality monitoring of North Seawater

The samples of raw North seawater from Jacobahaven, the Netherlands were collected to monitor algal cell density, chlorophyll-a concentration, and biopolymer concentration for a period from February to December 2016. As illustrated in Figure 5.10, the highest algal cell concentration (872 cells/mL), chlorophyll-a concentration (13 µg/L) and biopolymer concentration (0.16 mg-C/L) were recorded during April-May, 2016. The continuous monitoring of North seawater showed a decline in algal cell density. The results of monitoring indicates that the duration of the measured highest concentration (algal cell density, chlorophyll-a, and biopolymer concentration) lasts only for few days and shows no algal blooms during the period when the pilot testing was performed. Algal blooms are defined when the algal cell density ranged from 1000 cells/mL to 600,000 cells/mL

(Villacorte et al., 2015c). Therefore, to perform the further experiments in the pilot testing, an artificial algal bloom was simulated by culturing the marine algae *Chaetoceros affinis* as described in section 5.2.1. The feed solution needed for UF experiments was prepared by mixing the algal organic matter (AOM) harvested from algal culture and the raw North Seawater. The characterization of UF feed water is described in the section 5.3.3.2.

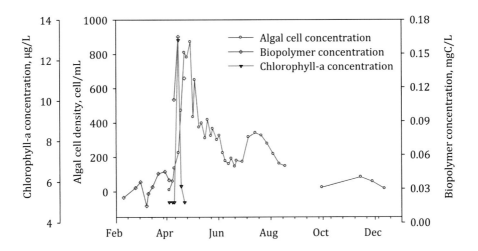

Figure 5.10: Monitoring of algal cell, chlorophyll-a, and biopolymer concentration of North seawater during a period from February to December 2016

5.3.3.2 Characterization of UF feed water

The average characteristic of the algal culture, raw North seawater, and ultrafiltration feed water during the period of pilot testing are summarized in Table 5.4. As illustrated in Table 5.4, the measured average algal cell density in algal culture was 15,000 ± 2,500 cells/mL, and chlorophyll-a concentration was 83±38, which is greater than the recommended threshold values beyond which algal blooms occur (Villacorte et al., 2015c). Likewise, the measured biopolymer concentration was approximately 1.76 mg-C/L, which is much higher than the reported value during algal blooms of 2009 in the North seawater (0.5 mg-biopolymer-C/L) (Villacorte, 2014). All these measured parameters indicate the situation of algal blooms. The algal organic matter (AOM) harvested from algal bags were 5 times diluted with the North seawater to have a final AOM concentration of approximately 0.5 mg-biopolymer-C/L in UF feed tank. The TOC difference in UF feed tank after dilution of

AOM with North seawater might be the contribution from EDTA added in algal bags (3.38 mg/L) in every 2 days as well as from carbon released by bacteria or external contamination. Moreover, the same EDTA concentration in the algal bag was maintained throughout the experiments.

Table 5.4: Characteristic of the algal culture (algae + AOM), seawater, and UF feed (AOM + seawater)

Parameters	Units	Algal bags (Algae + AOM)	Seawater	UF feed (AOM+ seawater)
Algal cell density	cells/mL	$15,000 \pm 2,500$	400	-
Chlorophyll-a concentration	µg/L	83 ± 38	BDL	BDL
Biopolymer concentration	mg-C/L	1.76	0.17	0.5 ± 0.1
TOC	mg/L	8.25	2.1	3.90
Modified fouling index (MFI-UF$_{10kDa}$)	s/L^2	n.m	6,000	15,610
pH		8.5	8.1	7.8 - 8.0
Temperature	°C	n.m	13 -14	18 - 23

5.3.3.3 Hydraulic performance of UF membranes

The fouling rate development in ultrafiltration membranes (10 kDa and 150 kDa) during multi-filtration cycles experiment was assessed. The AOM concentration of feed solution used during the experiment was approximately 0.5 mg-biopolymer-C/L. Both UFs were operated under the similar operating conditions as described in section 5.2.3.1. As illustrated in Figure 5.11a, the tight UF (10 kDa) showed 0.6 bar of transmembrane pressure (TMP) development during 100 hours of filtration, while it was about 0.3 bar in case of standard UF (150 kDa) membrane. The calculated non-backwashable fouling rates development after each succeeding CEB cycles were also found approximately 1.5 times higher for 10 kDa UF compared to 150 kDa UF (Figure 5.11b). The consequences of which might be the reduction in the surface porosity of 10 kDa UF membrane due to complete pore blocking of some pores. This phenomenon may increase the localized flux on the membrane, which eventually increases the ΔP, as pressure drop through cake layer is directly proportional to the square of filtration flux. The surface porosity of 10 kDa UF is about 0.9 % and that of 150 kDa UF is about 6.3 % (Tabatabai, 2014). The calculation of surface porosity was based on the analysis of F-ESEM image of UF membranes using AutoCAD with an assumption that all the visible pores as spherical in shape. The higher surface porosity of 150 kDa UF resulted for the lower non-backwashable fouling resistance development and lower TMP. It is expected that the cake/gel layer formed on the membrane surface be better removed by backwashing/CEB in a membrane with higher surface porosity.

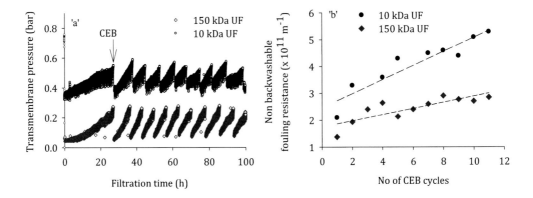

Figure 5.11: (a) TMP development over filtration hours and (b) Non-backwashable fouling resistance development, after each CEB cycles, when AOM obtained from Chaetoceros affinis (0.5 mg-biopolymer-C/L was fed through 10 kDa UF and 150 kDa UF membrane

5.3.3.4   Permeate quality assessment of hollow fiber UFs

The permeate quality of two hollow fiber UF membranes was assessed. Feed and permeate samples were collected every week and analyzed in terms of biopolymer concentration (using LC-OCD) and bacterial regrowth potential (using flow cytometry).

Biopolymer rejection

The average biopolymer concentration of UF feed was approximately 0.5 mg-C/L. This solution was fed to UF operated at a filtration flux of 40 L/m²/h, backwash with UF permeates at a backwash flux of 250 L/m²/h. The LC-OCD analysis of the collected permeate before the start of CEB cycles showed the biopolymer removal efficiency of approximately 90 % and 75 %, respectively for tight UF (10 kDa) and standard UF (150 kDa).

Reduction in bacterial regrowth potential

The bacterial regrowth potential of the collected feed and permeate samples of 10 kDa, and 150 kDa hollow fiber UF membranes were measured according to the protocol described in section 5.2.4.2. As presented in Figure 5.12a, the regrowth of inoculated natural consortium of marine bacteria in two samples (AOM and UF feed) showed a lag phase of 24 hours followed by an exponential growth phase until day 3 and reached a stationary or death phase. Likewise, other two samples, i.e., permeate of both UF membranes showed a lag phase of 48 hours followed by an exponential growth phase until day 4 and reached stationary/death phase. To better compare, the net live bacterial regrowth was calculated based on the difference between the maximum and initial concentration of bacteria in the incubated samples (Figure 5.12b). The result revealed that the bacterial reduction potential was 87 % and 89 %, respectively for 150 kDa UF and 10 kDa UF, which showed no remarkable difference between two UF permeates in terms of bacterial regrowth potential. It could be attributed to the contribution of low molecular weight organics that passes through both UF membranes.

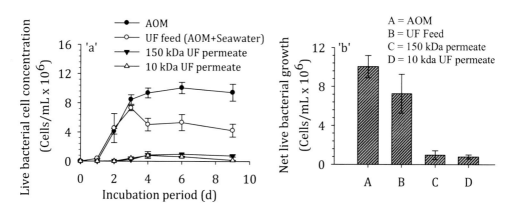

Figure 5.12: (a) Bacterial regrowth potential measured in feed and permeate of UF hollow fiber (10 kDa and 150 kDa) membranes, and (b) Calculated net live bacterial regrowth in feed and permeate of UF membranes

5.3.3.5   Pressure drop (ΔP) development in MFS cells

Three MFS monitors were operated in parallel with different feed solution such as; i) UF feed (MFS 1), ii) permeate of 150 kDa UF (MFS 2), and iii) permeate tight UF (10 kDa) (MFS 3). Within the research period of 15 days, no substantial increase in pressure drop was observed in the monitors fed with permeate of 150 kDa and 10 kDa UF, while a rapid increase in pressure drop was observed in MFS monitor fed with UF feed water. As illustrated in Figure 5.13a and b, the net increase in pressure drop was 1,146 mbar, 49 mbar, and 17 mbar, respectively in monitors fed with UF feed, 150 kDa UF permeate and 10 kDa UF permeate. After 15 days of operation, all monitors were opened for visual observation, and the biomass parameters such as ATP and TOC were measured. Biomass accumulation in the monitor fed with 10 kDa UF permeate was about 860 pg ATP/cm^2, which was approximately 2 and 5 times lower than in monitor fed with 150 kDa UF permeate and UF feed, respectively (Figure 5.13c). The measure ATP concentration in the MFS fed with 10 kDa UF permeate was less than the threshold (< 1,000 pg ATP/cm^2) recommended by (Vrouwenvelder et al., 2010) below which no pressure drop increased was observed. This indicates the potential of tight UF (10 kDa) towards delaying the occurrence of biofouling in SWRO membranes. However, a more extended period of operation is necessary to have a clear difference between the two molecular weight cut-off ultrafiltration membranes.

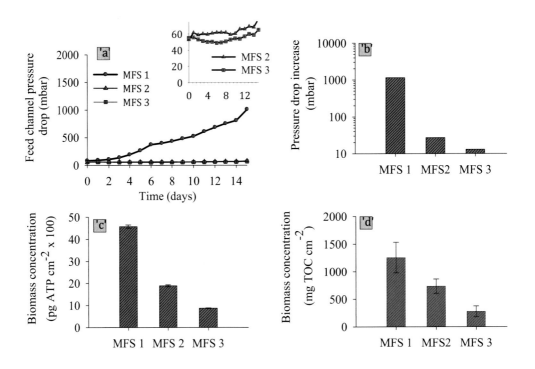

Figure 5.13: (a) Feed channel pressure drop in MFS monitors over time, (b) Increase in pressure drop in three MFS monitors, (c) and (d) biomass concentration measured in terms of ATP and TOC after 15 days of operation of MFS monitor fed with UF feed solution (MFS 1), permeate of standard UF (150 kDa) (MFS 2) and permeate of tight UF (10kDa) (MFS 3)

5.4   Conclusions

• The rejection of algal biopolymer produced by *Chaetocerous affinis* was 3 to 4 times higher with tight UF (10 kDa) compared to the high molecular weight cut off MF/UF membranes. The lower biopolymer concentration in permeate of tight UF coincided with the lower bacterial regrowth potential. This suggested that the tight UF has the potential in delaying the onset of organic/biological fouling in SWRO feed water during algal blooms.

• The biopolymer fraction of AOM was linearly correlated (R^2 = 0.88) with net bacterial regrowth. The LC-OCD analyses performed for feed, and permeate samples of 0.1 μm, 100 kDa, and 30 kDa and 10 kDa at 0 and 27 days of incubation suggested that bacteria are degrading the larger biopolymers more easily than, the smaller ones.

• No substantial difference was observed on the measured net bacterial regrowth in permeate of tight UF (10 kDa) and standard UF (150 kDa) in pilot scale testing. It could be attributed due to the contribution of low molecular weight organics that passes through both UF membranes.

• No substantial increase in feed channel pressure drop was observed during 15 days when permeate of 150 kDa and 10 kDa UF were fed into MFS monitors at a cross flow velocity of 0.2 m/s.

• The result of the membrane autopsy showed that the biomass accumulated in the MFS monitor fed with 10 kDa UF permeate was about 860 pg ATP/cm^2, which was 2 times lower than in monitor fed with 150 kDa UF permeate. The measure ATP concentration in the MFS fed with 10 kDa UF permeate was < 1,000 pg ATP/cm^2 below which no pressure drop increased was observed (Vrouwenvelder et al., 2010). This indicates the potential of tight UF (10 kDa) towards delaying the occurrence of biofouling in SWRO membranes. Nevertheless, it is still important to improve the backwashability performance of the tight UF membrane for the better future application.

5.5   Acknowledgements

Thanks to Marco Dubbeldam and Bernd van Broekhoven from Zeeschelp for algal culture, Tom Spanier, Henry Hamberg, Sander Brinks from Pentair X-flow for technical support in a pilot plant, Almotasembellah Abushaban for ATP measurement.

5.6   References

Babel, S. and Takizawa, S. (2010) Microfiltration membrane fouling and cake behavior during algal filtration. Desalination 261(1–2), 46-51.

Bar-Zeev, E., Berman-Frank, I., Girshevitz, O. and Berman, T. (2012) Revised paradigm of aquatic biofilm formation facilitated by microgel transparent exopolymer particles. Proc Natl Acad Sci U S A 109(23), 9119-9124.

Berman, T. and Holenberg, M. (2005) Don't fall foul of biofilm through high TEP levels. Filtration & Separation 42(4), 30-32.

Boerlage, Ś.F.E., Kennedy, M.D., Aniye, M.p., Abogrean, E.M., El-Hodali, D.E.Y., Tarawneh, Z.S. and Schippers, J.C. (2000) Modified Fouling Indexultrafiltration to compare pretreatment processes of reverse osmosis feedwater. Desalination 131(1), 201-214.

Caron, D.A., Garneau, M.-È., Seubert, E., Howard, M.D.A., Darjany, L., Schnetzer, A., Cetinić, I., Filteau, G., Lauri, P., Jones, B. and Trussell, S. (2010) Harmful algae and their potential impacts on desalination operations off southern California. Water Research 44(2), 385-416.

Flemming, H.C. (2002) Biofouling in water systems--cases, causes and countermeasures. Appl Microbiol Biotechnol 59(6), 629-640.

Flemming, H.C. (2011) Biofilm highlights, Springer - Verlag Berlin Heidelberg.

Guastalli, A.R., Simon, F.X., Penru, Y., de Kerchove, A., Llorens, J. and Baig, S. (2013) Comparison of DMF and UF pre-treatments for particulate material and dissolved organic matter removal in SWRO desalination. Desalination 322, 144-150.

Huber, S.A., Balz, A., Abert, M. and Pronk, W. (2011) Characterisation of aquatic humic and non-humic matter with size-exclusion chromatography – organic carbon detection – organic nitrogen detection (LC-OCD-OND). Water Research 45(2), 879-885.

Kim, K.J., Fane, A.G., Ben Aim, R., Liu, M.G., Jonsson, G., Tessaro, I.C., Broek, A.P. and Bargeman, D. (1994) A comparative study of techniques used for porous membrane characterization: pore characterization. Journal of Membrane Science 87(1), 35-46.

Matin, A., Khan, Z., Zaidi, S.M.J. and Boyce, M.C. (2011) Biofouling in reverse osmosis membranes for seawater desalination: Phenomena and prevention. Desalination 281, 1-16.

Mopper, K., Zhou, J., Sri Ramana, K., Passow, U., Dam, H.G. and Drapeau, D.T. (1995) The role of surface-active carbohydrates in the flocculation of a diatom bloom in a mesocosm. Deep Sea Research Part II: Topical Studies in Oceanography 42(1), 47-73.

Myklestad, S. and Haug, A. (1972) Production of carbohydrates by the marine diatom Chaetoceros affinis var. willei (Gran) Hustedt. I. Effect of the concentration of nutrients in the culture medium. Journal of Experimental Marine Biology and Ecology 9(2), 125-136.

Naidu, G., Jeong, S., Vigneswaran, S. and Rice, S.A. (2013) Microbial activity in biofilter used as a pretreatment for seawater desalination. Desalination 309, 254-260.

Pankratz, T. (2008) Red tides close desal plants. Water Desalination Report 44 (1).

Passow, U. (2002) Transparent exopolymer particles (TEP) in aquatic environments. Progress in Oceanography 55(3–4), 287-333.

Petry, M., Sanz, M.A., Langlais, C., Bonnelye, V., Durand, J.-P., Guevara, D., Nardes, W.M. and Saemi, C.H. (2007) The El Coloso (Chile) reverse osmosis plant. Desalination 203(1–3), 141-152.

Radu, A.I., Vrouwenvelder, J.S., van Loosdrecht, M.C.M. and Picioreanu, C. (2012) Effect of flow velocity, substrate concentration and hydraulic cleaning on biofouling of reverse osmosis feed channels. Chemical Engineering Journal 188, 30-39.

Reddy, V. (2009) Red Tide in the Arabian Gulf. MEDRC Watermark 40(3).

Salinas-Rodriguez, S.G., Amy, G.L., Schippers, J.C. and Kennedy, M.D. (2015) The Modified Fouling Index Ultrafiltration constant flux for assessing particulate/colloidal fouling of RO systems. Desalination 365, 79-91.

Salinas - Rodriguez, S.G., Kennedy, M.D., Schippers, J.C. and Amy, G.L. (2009) Organic foulants in estuarine and bay sources for seawater reverse osmosis - Comparing pre-treatment processes with respect to foulant reductions. Desalination and Water Treatment 9, 155-164.

Tabatabai, S.A.A. (2014) Coagulation and ultrafiltration in seawater reverse osmosis pretreatment, Ph.D. thesis UNESCO IHE.

Tabatabai, S.A.A., Schippers, J.C. and Kennedy, M.D. (2014) Effect of coagulation on fouling potential and removal of algal organic matter in ultrafiltration pretreatment to seawater reverse osmosis. Water Research 59, 283-294.

Uta, P. (2000) Formation of transparent exopolymer particles, TEP, from the dissolved precursor material. Marine Ecology Progress Series 192, 1-11.

Villacorte, L.O. (2014) Algal blooms and membrane-based desalination technology, Ph.D. thesis, UNESCO-IHE/TUDelft, Delft.

Villacorte, L.O., Ekowati, Y., Calix-Ponce, H.N., Schippers, J.C., Amy, G.L. and Kennedy, M.D. (2015a) Improved method for measuring transparent exopolymer particles (TEP) and their precursors in fresh and saline water. Water Research 70, 300-312.

Villacorte, L.O., Ekowati, Y., Winters, H., Amy, G., Schippers, J.C. and Kennedy, M.D. (2015b) MF/UF rejection and fouling potential of algal organic matter from bloom-forming marine and freshwater algae. Desalination 367, 1-10.

Villacorte, L.O., Kennedy, M.D., Amy, G.L. and Schippers, J.C. (2009) The fate of Transparent Exopolymer Particles (TEP) in integrated membrane systems: Removal through pre-treatment processes and deposition on reverse osmosis membranes. Water Research 43(20), 5039-5052.

Villacorte, L.O., Tabatabai, S.A.A., Dhakal, N., Amy, G., Schippers, J.C. and Kennedy, M.D. (2015c) Algal blooms: an emerging threat to seawater reverse osmosis desalination. Desalination and Water Treatment 55(10), 2601-2611.

Vrouwenvelder, H. (2009) Biofouling of spiral wound membrane system. Ph D thesis, Delft University of Technology, Delft.

Vrouwenvelder, J.S., Beyer, F., Dahmani, K., Hasan, N., Galjaard, G., Kruithof, J.C. and Van Loosdrecht, M.C.M. (2010) Phosphate limitation to control biofouling. Water Research 44(11), 3454-3466.

Vrouwenvelder, J.S., van Paassen, J.A.M., Wessels, L.P., van Dam, A.F. and Bakker, S.M. (2006) The Membrane Fouling Simulator: A practical tool for fouling prediction and control. Journal of Membrane Science 281(1–2), 316-324.

Zhang, X., Fan, L. and Roddick, F.A. (2015) Effect of feedwater pre-treatment using UV/H2O2 for mitigating the fouling of a ceramic MF membrane caused by soluble algal organic matter. Journal of Membrane Science 493, 683-689.

Annexes: Supporting information

Supplementary – S1: Artificial seawater preparation and culture media

Supplementary Table S.4.1: Inorganic ion composition of model artificial seawater (ASW)

Inorganic Ions	Concentration (g/L)
Chlorine (Cl^-)	18.85
Sodium (Na^+)	10.75
Sulphate (SO_4^{-2})	2.69
Magnesium (Mg^{2+})	1.17
Calcium (Ca^{2+})	0.30
Potassium (K^+)	0.38
Hydrogen Carbonate (HCO_3^-)	0.15
Total dissolved solids (TDS)	34.29

Supplementary Table S.4.2: Guillard's medium for marine diatoms (www.ccap.ac.uk)

Compound	Concentration (mL/1 L of SSW)
$NaNO_3$	1 (75g/L)
$NaH_2PO_4.2H_2O$	1 (5.65g/L)
Trace elements*	1
Vitamin mix**	1
$Na_2SiO_3.9H_2O$***	1
$NaHCO_3$ (Buffer solution)	4

* Trace elements (Chelated)**per litre**
NA_2EDTA-4.36 g
$FeCl_3.6H_2O$-3.15 g
$CuSO_4.5H_2O$-0.01 g
$ZnSO_4.7H_2O$-0.022 g
$CoCl_2.6H_2O$-0.01 g
$MnCl_2.4H_2O$-0.18 g
$Na_2MoO_4.2H_2O$-0.006 g
** Vitamin mix
Cyanocobalamin (Vitamin B_{12}) -0.0005 g
Thiamine HCl (Vitamin B_1)-0.1 g
Biotin-0.0005 g

*** Stir while adding

Supplementary – S2: Reproducibility test of MFS flow cells

In order to verify if the hydraulic conditions of the RO flow cells are reproducible and closely simulate the hydraulic conditions in SWRO, various lab tests were performed. The pressure drop across the MFS cells was recorded at various cross flow velocities ranging from 0.2 to 0.4 m/s. The experiment was performed with clean membranes/spacers and fed with artificial seawater. The results were compared with theoretical pressure drop calculation.

Theoretical pressure drop calculation

The theoretical pressure drop calculations were based on the Darcy–Weisbach equation for flat channels (Equation 1). For spacer filled channels, the friction coefficient (f) and hydraulic diameter (d_h) equations used by Schock and Miquel (1987) for spiral wound membranes were adopted.

$$\Delta P = \frac{f \rho v^2}{200} \frac{L_m}{d_h}$$

For spacer- filled channels

$$f = 6.23 R_e^{-0.3}$$

$$d_h = \frac{4\varepsilon}{2\frac{w+h}{wh}+(1-\varepsilon)S}$$

$$R_e = \frac{\rho v d_h}{\eta}$$

Where,

ΔP = pressure drop through the MFS cell (mbar)
f = friction coefficient (-)
v= cross-flow velocity (m/s)
ρ = water density, 1030 kg/ m³
L_m = the length of the membrane, 0.02 m
d_h = hydraulic diameter (m)
ε = spacer porosity; 0.89 (Vrouwenvelder, 2009)
S = specific surface of the membrane, 11600 m²/m³ (Vrouwenvelder, 2009)
w = width of spacer, 0.01m
h= height of spacer, 0.000787 m
η= dynamic viscosity, 0.001002 Pa.s
Re = Reynolds number (-)

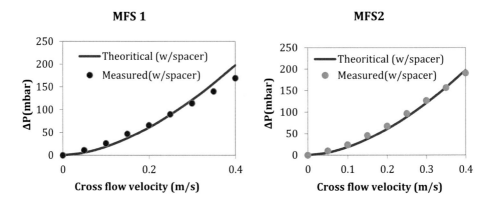

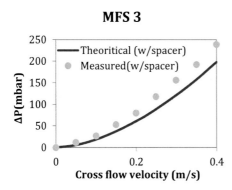

Supplementary Figure S.5.1: Reproducibility of three MFS cells

Supplementary – S3: Challenges during the operation of pilot plants

The pilot plant operation had several challenges during the Ph.D. period

i. Frequent shut down of the ultrafiltration operation

The operation of UF was frequently stopped which resulted in no permeate production. This had consequences for the operation of downstream Membrane Fouling Simulator (MFS). The capacity of UF permeates tank was 20 L, which was enough for 2.5 hours operation of MFS at a cross flow velocity of 0.2 m/s considering the dead volume needed for a startup the UF operation.

ii. Leaking from Amiad strainer

During the pilot plant operation, Amiad strainer had a leakage problem. The Amiad filter was also damaged many times mainly by the deposition of silt passing through the intake pump.

iii. Location of pilot

Physical access to the pilot plant located in Jacobahaven, the Netherlands was an issue especially for those who do not have a driving license.

iv. Algal culture

The continuous production of algal organic matter needed for a long-term operation of UF and MFS was a challenge during the pilot operation in both time and cost.

Supplementary Figure S.5.2: Three MFS cells on operation fed with A) 10 kDa UF permeate, B) 150 kDa UF permeate, and C) UF feed

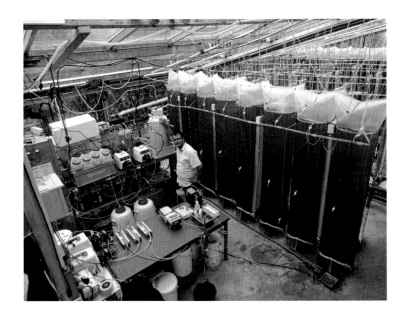

Supplementary Figure S.5.3: General overview of a pilot plant in Zeeland, Jacobahaven, the Netherlands. The large plastic bags are the algal culture of marine diatom chaetoceros affinis

6

Phosphate removal in seawater reverse osmosis feed water: An option to control biofouling

Contents

This chapter is based on:

Dhakal, N., Salinas Rodriguez, S.G., Schippers, J.C. and Kennedy, M.D. (2017), Phosphate removal in seawater reverse osmosis feed water: An option to control biofouling. *In preparation for Desalination*

Abstract

This chapter aimed to evaluate the efficiency of phosphate adsorbent coated with iron hydroxide group combined with UF towards delaying the onset of organic/biological fouling in SWRO feed water during algal blooms. Hereafter, the phosphate adsorbent is referred as phosphate removal technology (PRT™). A proof of principle experiments using PRT™ combined with UF membranes was tested using algal organic matter (AOM) harvested from cultured Chaetoceros affinis as a feed solution in both laboratory and pilot studies. In both cases, the feed and permeate samples were analyzed in terms of phosphate removal, biopolymer concentration removal, and reduction in bacterial regrowth potential. Furthermore, in pilot studies, biofouling experiments were performed using membrane-fouling simulators (MFS) to simulate biofouling in spacer-filled RO membrane channels when fed with permeate.

The result demonstrated that the application of PRT™ reduced the phosphate concentration to a level of 4 - 5 µg PO₄-P/L, which is approximately the removal efficiency of 98 %. The measured absolute phosphate concentration in permeate of PRT™ was similar to the detection limit of the Skalar San++ analyzer. The lower phosphate in permeate of the PRT™ coincided with the lower bacterial regrowth potential. The addition of 10 µg PO₄ – P/L in permeate of PRT™ showed the higher bacterial regrowth potential, which demonstrate the role of phosphate removal towards delaying the onset of biofouling in SWRO feed water. The biofouling experiments performed using MFS showed no increase in feed channel pressure drop for 21 days when fed with permeate of UF + PRT™. While there was a pressure drop increase of 500 mbar in MFS fed with permeate of UF alone. The result of membrane autopsy showed that the biomass accumulation was below the detection limit in the MFS fed with permeate of UF + PRT™, while it was about 6,000 pg ATP/cm² in the MFS fed with permeate of UF alone. The higher biomass accumulations in MFS fed with permeate of tight UF alone could be attributed to the passage of low molecular weight (LMW) organic and dissolved phosphate through UF. The possible contribution of LMW was demonstrated by the observed linear growth (R² = 0.65) between the EDTA concentrations and the net bacterial regrowth. Overall, the removal of phosphate by the application of PRT™ combined with tight UF (10 kDa) restricted the biomass growth and thus delay the onset of organic/biological fouling in SWRO feed water during algal blooms. Moreover, a more extended period of experiments is needed for further verification.

Keywords: *Phosphate removal technology, tight ultrafiltration, algal blooms, algal organic matter, seawater reverse osmosis*

6.1 Introduction

Seawater reverse osmosis (SWRO) is currently the leading and a fast-growing membrane-based desalination technology worldwide. A major problem identified during the operation of SWRO is the membrane fouling and more specifically biofouling (Baker et al., 1998, Flemming et al., 1997, Nguyen et al., 2012, van Loosdrecht et al., 2012). Biofouling refers to the operational problem associated with the accumulation and growth of the microorganism and the extracellular polymeric substances (EPS) they produced in the RO membranes. The processes involved in the occurrence of biofouling in RO membranes are categorized into four stages (Flemming et al., 1988) namely;

Surface conditioning: it refers to the formation of *"conditioning film"* when the adsorptions of macromolecules (humic substances, lipopolysaccharides, and other product of microbial turnover) start on the membrane.

Bacterial adhesion: the initial adhesion of the fast adhering bacterial cells from the raw water occurs on the conditioned surface of the membrane.

Bacterial colonization, growth, and biofilm formation: the primary colonization starts when the bacteria that primarily adhere to the membrane surface grow at the expense of available nutrients. The number of bacterial colonies and the production of EPS increases when more bacteria of different species adhere to the membrane surface, which leads to the formation of slime layer and is known as a biofilm.

Biofouling: biofouling occurs when the thickness of the slime layer (biofilm) or the hydraulic resistance increases beyond an operationally defined threshold of interference. The operational definition of biofouling in RO is the increase in the feed channel pressure drop or decrease in normalized flux by > 15 %.

The occurrence of biofouling in SWRO membranes results in i) a decrease of membrane permeability, ii) increase in the operating pressure, iii) increase in the frequency of chemical cleaning and iv) increase in the frequency of membrane replacement (Matin et al., 2011, Radu et al., 2012). Several factors play a role in the occurrence of biofouling in SWRO membranes namely; cross-flow velocity, feed spacer and biodegradable nutrients available in the feed water (Dreszer et al., 2014, Vrouwenvelder et al., 2009).

Flemming (2011) suggested the most promising strategy for biofouling control is to limit the bacterial and the nutrient concentration (C, N, and P) from the RO feed water using a reliable pretreatment system (Flemming, 2011). In most cases, organic carbon is considered to be the limiting nutrient for microbial growth (Kooij et al., 1982). The existing pre-treatment systems such as biologically activated carbon filtration or slow sand filtration, ultrafiltration are capable of reducing the concentration of biodegradable organic carbon from RO feed water to some extent (Vrouwenvelder et al., 2010). Moreover, the application of tight UF (molecular weight cut off 10 kDa) was found capable of rejecting higher concentration of organic carbon but allows the passage of dissolved phosphate (see chapter 5 of this thesis).

It has been demonstrated that limiting phosphate from RO feed water can restrict biological growth even in the presence of high concentrations of other nutrients (Vrouwenvelder et al., 2010). Jacobson et al. (2009) also suggested that phosphate limitation can be a solution to control biofouling occurrence in SWRO (Jacobson et al., 2009). The required phosphorous levels are approximately 10 times lower than the carbon level as indicated by the molar ratio of carbon (C), nitrogen (N) and phosphorous (P), i.e.,~100:20:10 required for microbial growth (Vrouwenvelder et al., 2010). In nature, phosphorous is present in the form of orthophosphate (H_3PO_4, $H_2PO_4^-$, HPO_4^{2-}), inorganic polyphosphates and dissolved organic phosphorus (Holtan et al., 1988, Vrouwenvelder et al., 2010). Among them, orthophosphate is the most readily available form of phosphorus (Maher et al., 1998). In this study, orthophosphate is referred to as phosphate.

The phosphate concentration in seawater is usually less than 33 μg PO_4 - P/L (Garcia, 2006) with an average of 20 μg PO_4 - P/L (OZ REEF, 2007) as cited by (Vrouwenvelder et al., 2010). Jacobson et al, (2009) reported that phosphate concentration below 1.5 μg PO_4 -P/L might limit/eliminate biofouling in reverse osmosis membranes (Jacobson et al., 2009). Likewise, the study conducted by Vrouwenvelder et al, (2010) also found that phosphate concentration ~ 0.3 μg PO_4-P/L in the feed water restricted the increase in pressure drop in membrane fouling simulator (MFS). However, it is a challenge to remove the phosphate concentration from RO feed water to such a low level. In practice, chemical treatment using the precipitants such as lime, alum, and ferric chloride is common practice to remove phosphate(Ugurlu et al., 1998). Moreover, the currently applied NF and RO membranes are sensitive to free chlorine

(Vrouwenvelder et al., 2010). Sevcenco et al. (2015) reported that the currently available techniques that can remove phosphate from the water sample are not sustainable (Sevcenco et al., 2015), which highlighted the importance of alternative phosphate removal technology to be sustainable and environmentally friendly.

In this work, we aimed to evaluate the efficiency of bio-based weakly anion exchange resin-coated with iron hydroxide group towards phosphate removal and biofouling control in SWRO membranes during an algal bloom. The adsorbent tested in this study was developed by BiAqua in the Netherlands and is a regenerable material. Hereafter, the phosphate adsorbent is referred as phosphate removal technology (PRT™). Biaqua claimed that PRT™ has the potential to reduce the phosphate levels below 1 - 2 µg/L. The goal of this study was to verify experimentally whether or not the application of PRT™ combined with ultrafiltration be a suitable pre-treatment to eliminate or delay the biofouling in SWRO membranes during an algal bloom. The artificial algal bloom condition was created by culturing marine bloom-forming algae *"Chaetoceros affinis."* The specific objectives are;

i. to characterize the PRT™ adsorbent using scanning electron microscope (SEM), Fourier transform infrared spectroscopy (FTIR), X-ray diffraction (XRD), and inductively coupled plasma mass spectrometry (ICP - MS)

ii. to determine the phosphate removal efficiency of the PRT™ when treated with algal bloom impacted water

iii. to check the effectiveness of PRT™ combined with ultrafiltration membrane in delaying the onset of biofouling in SWRO feed water during algal blooms

6.2   Material and methods

6.2.1 Algal culture, AOM extraction, and AOM characterization

A common bloom-forming marine diatom *Chaetoceros affinis* was grown to represent the algal bloom situation in seawater. The algal strain (CCAP 1010/27) was purchased from the Culture Collection of Algae and Protozoa (SAMS, Scotland). The algal strain was inoculated in sterilized synthetic seawater (TDS = 35 g/L, pH = 8.0 ± 0.3) and spiked with nutrients and trace elements based on Guillard's (f/2 + Si) medium necessary for them to grow rapidly and simulate an algal bloom. After 13 -14 days under controlled light (12 h light: 12h dark) and continuous slow mixing at an ambient temperature of 20 °C, the algal cells were separated from the medium containing AOM by allowing the cells to settle for 24 hours. The supernatant was then extracted and filtered using 5 μm polycarbonate filter (Whatman Nuclepore) to remove the remaining algal cells in the suspension. The extracted algal organic matter (AOM) solution was stored at 5 °C and sent to DOC-Labor (Karlsruhe, Germany) and Wetsus, Leeuwarden for LC-OCD characterization. The extracted AOM solutions were used for the proof of principle experiments in the laboratory scale.

The algal production was scaled up for the demonstration experiments in a pilot plant, where a marine diatom *Chaetoceros affinis* was cultured in 250 L plastic bags (6 parallel bags) using 2 μm filtered and UV disinfected North Seawater as shown in Figure 6.1.

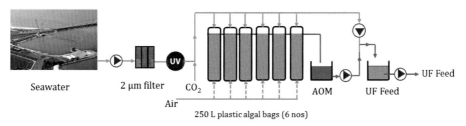

Figure 6.1: Algal culture, AOM production from marine diatom Chaetoceros affinis in pilot study

The detailed procedure for algal culture and harvesting is as explained in chapter 5 section 5.2.1. In short, the 2 μm filtered and UV disinfected North seawater was pumped at a flow rate of 1 L/h to each bag where nutrient (f/2 + Si) needed for rapid algal growth was periodically supplied. The solution was continuously mixed by supplying air from the bottom of the bag. The pH of the solution inside the bag was adjusted to a level of ~ 8.5 by adding CO_2. The cumulative AOM production at a rate of 6 - 7 L/h was cautiously harvested from the supernatant of the solution inside the bag. The liquid chromatography organic carbon detection (LC-OCD) analysis of the collected AOM, and North seawater was performed in Wetsus, Leeuwarden. The AOM and North seawater were mixed to have the final AOM concentration of 0.5 mg - biopolymer - C/L and were used as a feed solution during pilot testing.

6.2.2 Proof of principle in laboratory

An overview of the experimental approach in laboratory scale is presented in Figure 6.2. As illustrated in Figure 6.2, the AOM produced by *Chaetoceros affinis* was first filtered through hollow fiber UF membrane (molecular weight cut-off 150 kDa and 10 kDa), the permeate of which were again separately filtered through two different PRT™ column packed with phosphate adsorbent. The AOM used were pre-treated with 0.22 μm pre-filtration and two times diluted with autoclaved (121 °C, 20 minutes) artificial seawater to have the AOM concentration of ~ 0.97 mg – biopolymer - C/L. The details procedure and experimental conditions are explained below.

UF membrane preparation and AOM filtration

Two hollow fiber UF membranes pen modules (150 kDa and 10 kDa) were prepared using Pentair X-Flow capillary membrane fibers (6 numbers) with an internal diameter of 0.8 mm and effective filtration length of 30 cm. The effective surface area was 45 cm^2 ± 2 %. All fibers were potted inside 8 mm outer diameter polyethylene tubing (Festo, Germany) using polyurethane glue (Bison, The Netherlands) and left to dry for 24 hours. Before the filtration, the pen modules were soaked in hot water at 40 °C for 24 hours, removed all entrapped air inside and outside the membrane fibers. The clean water membrane resistance (R_m) was measured using MilliQ water at a flux of 50 L/m^2/h. The AOM as a feed solution was filtered through 10 kDa and 150 kDa hollow fiber UF modules at a filtration flux of 50 L/m^2/h. The filtration cycle of 20 minutes was used for both types of UFs followed by 45 seconds of backwashing at a rate of 125 L/m^2/h with UF permeate. The permeate of first filtration cycle was discarded, and then the total permeate volume of 500 mL weres collected. In addition, 500 mL of feed solution was also collected and analyzed biopolymer concentration, phosphate concentration, and bacterial regrowth potential.

Filtration of UF permeate through PRT™

A phosphate adsorbent (15 g) was transferred to a custom-made column (1.6 cm in diameter and 16 cm bed height) equipped with a metal filter at the bottom of the column. The column was connected to a gear pump. The system was intensively rinsed with milliQ water to remove any fines or pollutants from the adsorbent. The filtration of UF permeate was performed for about 30 minutes in a down-flow mode at a flow rate of 0.4 L/h calculated based on the empty bed contact time of 2 minutes. A 20 mL of the filtered volume was discarded before collecting 500 mL for analyzing biopolymer concentration, phosphate concentration, and bacterial regrowth potential.

Figure 6.2: Scheme of laboratory scale ultrafiltration and PRT™ filtration experiment

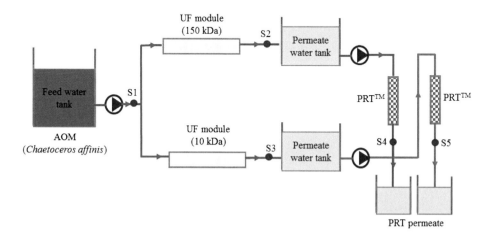

6.2.3 Investigation in a pilot plant

The effectiveness of hollow fiber polyethersulphone (PES) ultrafiltration membrane (molecular weight cut-off, 10 kDa) combined with PRT™ in reducing the biofouling in SWRO membranes during algal bloom were investigated in a pilot testing. The pilot plant was located in the estuarine region of South-Western Netherlands, Zeeland Province. The general scheme of the pilot plant is as shown in Figure 6.3. As illustrated in Figure 6.3, the pilot comprises of submerged open seawater intake (4 m depth), 50 μm Amiad strainer, a bench scale ultrafiltration systems and membrane fouling simulator (MFS). The raw seawater (40 L/h) after Amiad strainer was collected in a tank A and the remaining flow of raw seawater was drained to the sea. The collected raw seawater in the tank A was pumped using a peristaltic pump and was mixed with the AOM harvested from *Chaetoceros affinis*. The solutions were mixed hydraulically to have the final AOM concentration of 0.5 mg – biopolymer - C/L in the UF feed tank. The operational details of ultrafiltration are explained in section 6.2.3.1.

6.2.3.1 Ultrafiltration experiments

Hollow-fiber polyethersulphone (PES) UF membranes (molecular weight cut-off, 10 kDa) obtained from Pentair X-Flow were used for this experiment. Membrane modules were fabricated in Pentair X-Flow, Enschede, the Netherlands by potting 120 numbers of capillary fibers (internal diameter of 0.8 mm and 95 cm effective length) inside 1inch outer diameter polyethylene tubing. The effective membrane surface area was approximately 0.28 m².

The AOM that contains 320 µg PO₄ - P/L and 0.5 mg – biopolymer - C/L was fed to the ultrafiltration membrane. The detail operational conditions applied for UF filtration experiments are summarized in Table 6.1.

Table 6.1: Operational conditions for UF experiments

Parameters	Value
Filtration flux	40 L/m²/h
Filtration time	20 minutes
Backwash time	1 minute
Backwash flux	250 L/m²/h
CEB start when permeability	< 70 L/m²/h/bar
CEB flux	125 L/m²/h
CEB dosing time	2 minutes
Soaking time	10 minutes
Coagulation	No

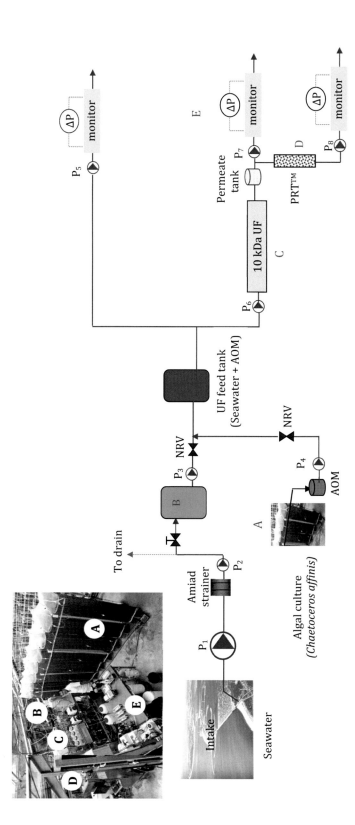

Figure 6.3: General scheme of the investigation pilot plant in this study. Abbreviations: P_1: Intake pump; P_2: Booster pump; P_3: Masterflex pump (Seawater dosing pump); P_4: AOM dosing pump; P_5 and P_6: UF feed pump; P_7, P_8 and P_9: Masterflex pump for membrane fouling simulator (MFS); NRV: Non-return valve

Picture showing a general outlook and the components of pilot plants (A) algal culture of marine diatom "Chaetoceros affinis" (B) raw seawater collection tank after 50 μm Amiad strainer, (C) A bench scale (1") ultrafiltration unit (10 kDa UF) (D) 4 cm diameter PRT™ column, and (E) Membrane Fouling Simulator

Prior to the AOM filtration, the virgin UF membranes were flushed with tap water to remove any preservative materials and any released materials from the membrane. All the entrapped air inside and outside the fibers of the membrane modules were also released.The clean water membrane resistance (R_m) of the UF (10 kDa) membrane were measured at the filtration flux (40 L/m²/h) using tap water. The principle how UF was operated in this study is presented in Figure 6..4. As illustrated in Figure 6.4, the ultrafiltration membrane was fed with AOM from both ends and inside-out mode, while permeate was drawn from one end of the module. The AOM filtration of 20 minutes was followed by 1 minute of backwashing with UF permeate at a flux of 250 L/m²/h. The transmembrane pressure (TMP) development in each filtration cycle was recorded using a pressure sensor. All the experiments were performed with no coagulation. The chemical enhanced backwash (CEB) of the UF membrane was performed using chemicals prepared onsite using tap water. The criterion for performing CEB was the decline of permeability of the membrane below 70 L/m²/h. The CEB was first performed with acid dosing (pH = 2 - 3), rinsed with UF permeate and soaked for 10 minutes. Secondly, the membranes were cleaned with a solution mixed with NaOH and NaOCl solution (pH = 11, chlorine concentration = 200 mg/L), rinsed again with UF permeate and soaked for 10 minutes.

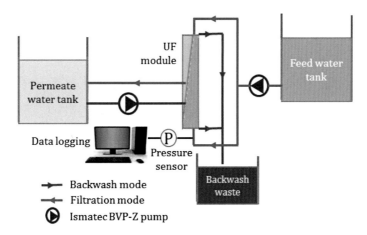

Figure 6.4: Scheme for ultrafiltration experiments

The permeate of ultrafiltration membrane was collected in a permeate tank and fed to the PRT™ column packed with bio-based weakly anion exchange resin-coated with iron hydroxide group. The design of PRT™ column and its operational criteria are explained in section 6.2.3.2.

6.2.3.2 Phosphate removal technology (PRT™)

PRT™ column used in this study consists of two fixed bed columns (4.4 cm in diameter). The feed flow rate to PRT™ was calculated as follows;

Column diameter	= 4.4 cm
Surface area	= 0.00152 m²
Bed Height	= 100 cm
Bed Volume	= 1,520.5 mL
Empty bed contact time (EBCT)	= 10 min
Velocity = Bed height /EBCT	= 6 m/h
Feed flow rate = Velocity x surface area	= 152.1 mL/min
	= 9.12 L/h

The selection of EBCT of 10 minutes was based on the capacity of the permeate production by ultrafiltration membrane after deducting the required volume of permeate for backwashing, CEB and needed dead volume in the tank for UF operation. The collected UF permeate in a 20 litres of buffer tank was pumped at a flow rate of 9.12 L/h using the multichannel peristaltic pump and fed to PRT™ in downflow mode. The formation of a channel in the filter bed was avoided by maintaining a 5 - 8 cm water column height on the top of the filter bed by adjusting the feed and permeate of the PRT™. The permeate of PRT™ was collected in a tank and fed to membrane fouling simulator (MFS) using the multichannel peristaltic pump as described in section 6.2.3.3.

6.2.3.3 Membrane Fouling Simulator (MFS)

Membrane fouling simulator (MFS) monitors with dimension 20 cm x 1 cm x 0.078 cm were used to simulate a biofouling in spacer filled RO membrane channels. The design specifications of MFS cells are according to described by (Vrouwenvelder et al., 2006). Three MFS experiments were run in parallel, i.e., i) MFS 1 fed directly with the UF feed water, ii) MFS 2 fed with permeate of tight UF (10 kDa) and iii) MFS 3 fed with permeate of tight UF (10 kDa) combined with PRT™.

The 4 inch RO membrane and spacer (TMD 810) purchased from Toray was used in MFS monitor. The virgin RO membrane module was opened and cut into coupons of dimension

20 cm x 1 cm using cutter at Wetsus, Leeuwarden. The membrane coupons were stored in 1 % sodium bisulphate solution for at least 24 hours and rinsed with water before placing them in MFS monitor. The orientation of spacer was made similar in all experiments. All three MFS monitors were operated without permeation at a cross flow velocity of 0.2 m/s, which is equivalent to a flow rate of 4.8 L/h. The feed flow was pumped using a multi-channels peristaltic master flex pump. A pressure drop transmitter (PMD70-2071/0, Endress + Hauser) was connected to monitor the feed channel pressure drop developed in MFS overtime. All the generated data at every 30 second were recorded in Ecograph T RSG35 data manager (Endress + Hauser).

6.2.3.4 Membrane autopsy

At the end of the experiment, the fouled membrane and spacer were cut into pieces with known dimensions. The samples of membrane and spacer were placed inside 50 mL clean tubes and filled with 40 mL of autoclaved (121 °C, 20 minutes) artificial seawater. The tubes were tightly covered, vortexed (Heidolph REAX 2000) for 10 seconds, sonicated (Branson 2510E-MT) for 5 minutes at a frequency of 42 kHz. The sonicated samples were analyzed by measuring adenosine triphosphate (ATP).

6.2.4 Characterization techniques

6.2.4.1 Feed and permeate characterization

6.2.4.1.1 Biopolymer concentration

The biopolymer concentration was quantified using the liquid chromatography organic carbon detection (LC-OCD) in DOC-Labor (Karlsruhe, Germany) and Wetsus facilities (Leeuwarden, the Netherlands). The LC-OCD analysis fractionates organic matters regarding biopolymers, humic substances, building blocks, low molecular weight (LMW) organic acids and neutrals. The procedure for organic carbon fractionation is described by (Huber et al., 2011). All the samples were pre-filtered through 0.45 μm Millipore filter before analysis.

6.2.4.1.2 Phosphate concentration

Phosphate analysis was performed using Skalar San++ analyzer (www.skalar.com) at the facility of WGML Anorganische analyze Rijkswaterstaat CIV), the Netherlands. The Skalar

San++ is a continuous flow analyzer. A sample was first mixed with molybdate reagent and ascorbic acid at a temperature of 37 °C. The added molybdate and the orthophosphate present in a sample forms a phosphor-molybdate complex in the acidic environment after reduction with ascorbic acid and in the presence of antimone. This gave a blue colored complex, which was measured at 880 nm using 50 mm cuvette and spectrophotometer.

6.2.4.1.3 Bacterial regrowth potential

The bacterial regrowth potential of feed and permeate samples were measured using the method described in Chapter 3 of this thesis. In short, the method involves the removal/inactivation of bacteria (live + dead) from the water sample, followed by re-inoculation with live natural consortium marine bacteria (10^4 cells/mL), incubation (30 °C), and monitoring of microbial growth using flow cytometry. The net live bacterial regrowth was calculated from the bacterial growth curve for the data interpretation.

6.2.4.2 Characterization of PRT™ adsorbent

6.2.4.2.1 Elemental analysis using ICP-MS

The presence of elements on the coated particles of the PRT™ adsorbents was performed after boiling the adsorbent in 65 % HNO_3 solutions. About 100 - 150 grams of PRT™ adsorbents were placed in a beaker and washed with demi water. The materials were dried at room temperature (20 °C). Approximately 0.5 g of dried materials were put in a 100 mL of the conical flask (in triplicate) and added with 40 mL of demi-water and 10 mL of 65 % HNO_3 to each flask. All the flasks were heated at 250 °C until approximately 10 mL of solution remain in the flask. The flasks were allowed to cool to room temperature, and all the materials were then collected and placed in 100 mL measuring flasks. Demi water was added to the measuring flasks until the 100 mL of the indicated mark. From the measuring flask, 5 mL of the solution were retrieved from the supernatant and diluted 20 times with demi water. The diluted samples were analyzed using the inductively coupled plasma mass spectrometry (ICP - MS).

6.2.4.2.2 Fourier transforms infrared spectroscopy (FTIR)

FTIR spectroscopy was applied to identify the functional groups presents on the coating particles of PRT™. Infrared (IR) spectra of the granular samples were recorded using Spectrum 100 FTIR spectrometer from PerkinElmer equipped with a Universal ATR

Sampling Accessory at the Aerospace Engineering Laboratory of the Delft University of Technology. Spectra were taken over 4 scans with a spectral range of 4,000 - 600 cm^{-1}. The peak-picking feature of the spectrum analysis software was used to identify significant peaks of interest.

6.2.4.2.3 X-ray diffraction (XRD)

XRD was applied to identify the phase of the PRT™ adsorbent. The sample was analyzed using Bruker advanced diffractometer, Bragg-Brentano geometry, and Lynxeye position sensitive detector at TU Delft. Data collection was carried out at room temperature using Cu Kα radiation in the 2θ region between 10 ° and 130 °, step size 0.034 degrees 2θ. Data evaluation was done with the Bruker software diffraction EVA version 4.2.

6.3   Results and discussion

6.3.1 Proof of principle in laboratory scale

The first hypothesis linking to the reduction of biofouling potential in SWRO is the removal of biopolymer and phosphate from the SWRO feed water. It is expected that limiting nutrients in the SWRO feed water; the biofilm growth can be prevented in the SWRO membranes surfaces. To evaluate these characteristics the reduction potential of UF - PRT™ in terms of biopolymer concentration, phosphate concentration, and bacterial regrowth potential were investigated as discussed in the following sections.

6.3.1.1 Rejection of biopolymer and phosphate by UF-PRT™

The rejection efficiency of UF - PRT™ in terms of biopolymer and phosphate concentration is presented in Figure 6.5. As illustrated in Figure 6.5a, ultrafiltration membranes (150 kDa and 10 kDa) were capable of rejecting the biopolymer concentration but allowed the passage of dissolved phosphate. The measured AOM concentration of the feed solution was 0.97 mg-biopolymer-C/L, of which 56 % was rejected by the UF (150 kDa) and 97 % by the UF (10 kDa). The results also illustrated that the subsequent addition of PRT™ after UF contributes for higher biopolymer rejection. The biopolymer rejection increased from 56 % to 74 % by UF (150 kDa) combined with PRT™. Nevertheless, no substantial difference on biopolymer rejection was observed in case of UF (10 kDa) + PRT™. The reported concentration of biopolymer is not the blank corrected. The blank (synthetic seawater) had a biopolymer concentration of 0.05 mg -C/L. The removal efficiency has increased by 3 - 4 % after the blank correction.

Likewise, Figure 6.5b shows the removal efficiency of dissolved phosphate by UF - PRT™. As illustrated, the measured phosphate concentration in the feed solution was approximately 320 µg PO₄ - P/L. This phosphate concentration is approximately 16 times higher than the average phosphate concentration measured in natural seawater, i.e., 20 µg PO₄-P/L (Vrouwenvelder et al., 2010). The PRT™ demonstrated the substantial reduction of phosphate concentration from 320 µg PO₄ - P/L to 6 µg PO₄ - P/L. Moreover, the phosphate concentration of 6 µg PO₄ - P/L measured in permeate of PRT™ is still much higher compared to the recommended threshold value beyond which biofouling in RO membranes would occur. For instance, Jacobson et al. (2009) reported that phosphate concentration below 1.5 µg PO₄-P/L might limit/eliminate biofouling in reverse osmosis membranes (Jacobson et al., 2009). Likewise, the study conducted by Vrouwenvelder et al., (2010) also suggested that phosphate concentration of ∼ 0.3 µg PO₄ - P/L in the feed water restricted the pressure drop increase in membrane fouling simulator (Vrouwenvelder et al., 2010).

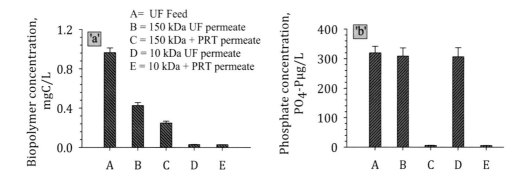

Figure 6.5: Biopolymer and phosphate concentration measured in UF feed (AOM), 150 kDa UF permeate, 150 kDa UF + PRT™, 10 kDa UF permeate and in permeate of 10 kDa + PRT™

The reported phosphate concentration of 6 µg PO₄ - P/L in permeate of PRT™ could due to the limit of detection (LOD) of the method applied. The claimed LOD for Skalar San⁺⁺ analyzer was 0.3 µg PO₄ - P/L. Moreover, the verification experiments performed with known standard samples prepared using synthetic seawater and spiked with different concentration of NaH₂PO₄ (0 - 100 µg PO₄ - P/L) showed a LOD of 5.5 µg PO₄ - P/L (Figure

6.6 inset). However, the linear correlation ($R^2 = 0.99$) observed between the expected PO_4 - P and the measured $PO_4 - P$ suggested the reliability of Skalar San[++] analyzer to measure phosphate in seawater sample.

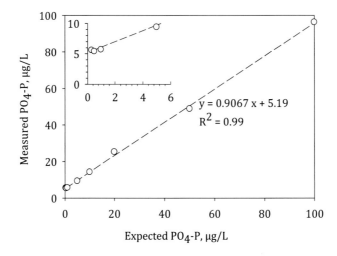

Figure 6.6: Accuracy and reliability of phosphate analyzer Skalar San++ while measuring phosphate in seawater sample

6.3.1.2 Bacterial regrowth potential

The regrowth potential of bacteria in feed and permeate of UF-PRT™ was investigated based on bacterial regrowth potential tests. All samples were pre-filtered with 0.22 µm and re-inoculated with a natural consortium of marine bacteria. The average concentration of inoculated bacteria in each sample was about 15,000 cells/mL. The samples were incubated at 30 °C and bacterial regrowth was monitored using flow cytometry for 33 days.

As illustrated in Figure 6.7, bacteria growing in all samples except in the UF feed showed a lag phase of 1 day followed by an exponential phase and a stationary phase. While no apparent lag phase was observed in UF feed and it might be due to the low frequency (24 hours) of sampling.

Despite having been inoculated with the same concentration of bacteria, the maximum bacterial regrowth was found different for different samples. To better compare, the net bacterial regrowth was calculated based on the difference between the maximum and initial concentration of bacteria (Table 6.2). As depicted, the net bacterial regrowth in the UF feed was approximately 20 million cells/mL. The measured net bacterial regrowth in permeate of 10 kDa UF was about 5 million cells/mL, which was 75 % lower than in the feed solution. While on the other hand, the net bacterial regrowth in permeate of 150 kDa UF was about 11.7 million cells/mL, which was about only 41 % lower than in the feed solution. The lower bacterial regrowth monitored in permeate of tight UF (10 kDa) compared to permeate of 150 kDa could be due to the higher biopolymer rejection by tight UF (refer to Figure 6.5a). The application of PRT™ combined with UF further reduced the net bacterial regrowth to 2.5 - 3.0 million cells/mL independent of the molecular weight cut-off of the UF membranes, which could mainly attribute to the phosphate removal by the PRT™ as mentioned in Figure 6.5b. The reported percentage of bacterial regrowth reduction potential was not the blank corrected. The blank (synthetic seawater) showed a bacterial regrowth of about 0.77 million cells/mL. The percentage of bacterial regrowth reduction potential has increased by 2 – 4 % after the blank correction.

Table 6.2: Percentage reduction of net bacterial regrowth potential after treatment through UF and PRT™

Filtration options	Net bacterial regrowth (10⁶ cells/mL)	Bacterial regrowth potential reduction (%)	
		Before blank correction	After blank correction
Feed solution	19.9	-	
150 kDa UF permeate	11.7	41	43
150 kDa UF + PRT™	3.0	85	88
10 kDa UF permeate	5.0	75	78
10 kDa UF + PRT™	2.5	87	91

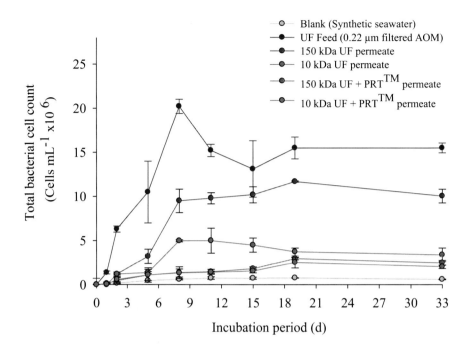

Figure 6.7: Growth curves of inoculated natural bacterial consortium in various samples such as blank (ASW), UF feed (0.22 µm filtered AOM), 150 kDa UF permeate, 150 kDa UF +PRT™ permeate, 10 kDa UF permeate, and 10 kDa UF + PRT™ permeate

This "proof of principle" experiment shows that the application of PRT™ combined with UF (10 kDa) substantially rejected the phosphate and biopolymer concentration which coincided with the lower bacterial regrowth potential, a surrogate parameter for biofouling. Moreover, the virgin PRT™ resin might also release the carbon, which can influence the bacterial regrowth. This hypothesis was tested by flushing the virgin PRT™ resin packed in a custom-made column with 1.6 cm diameter and 16 cm bed height. The flow rate of 6.4 mL/minute calculated based on empty bed contact time (EBCT) of 2 minutes was pumped into the column in down flow mode using a gear pump. The total organic carbon (TOC) of the feed and permeate samples were measured over time as shown in Figure 6.8. The result confirmed that the virgin PRT™ resin released the carbon, but it can be flushed out with carbon-free water in an approximately 120 minutes as illustrated in Figure 6.8.

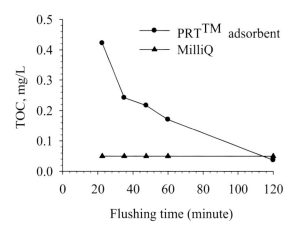

Figure 6.8: Measurement of TOC concentration released by Virgin PRT™ adsorbent over time when flushed with MilliQ water

6.3.2 Investigation in pilot scale the application of tight UF + PRT™

The potential of tight UF (10kDa) combined with PRT™ in eliminating/delaying the occurrence of biofouling in SWRO membranes during algal blooms was evaluated in a pilot scale experiments. The biofouling experiments were also performed using membrane fouling simulators (MFS) to simulate biofouling in spacer-filled RO membrane channels when fed with permeate of UF combined with PRT™. The description of the pilot plant is as described in section 6.2.3. The feed water used was the mixture of North seawater, and AOM harvested from *Chaetoceros affinis* as described in section 6.2.1. The results are discussed in the following sections.

6.3.2.1 Characterization of feed water

The average typical characteristic of the algal culture, raw North seawater and ultrafiltration feed water during the period of pilot testing are summarized in Table 6.3. As illustrated in Table 6.3, the measured average algal cell concentration in algal culture was 12,000 ± 3,500 cells/mL and chlorophyll-a concentration was 72 ± 38, which is higher than the recommended threshold values beyond which algal blooms occur (Villacorte et al., 2015b). Likewise, the measured biopolymer concentration was approximately 1.76 mg-C/L, which is much higher than the reported value during algal blooms of 2009 in the North seawater (0.5

mg - C/L) (Villacorte, 2014). All these measured parameters indicate the situation of algal blooms. The algal organic matter (AOM) harvested from algal bags were 5 times diluted with the North seawater to have a final AOM concentration of approximately 0.5 mg – biopolymer - C/L in UF feed tank. The TOC difference in UF feed tank after dilution of AOM with North seawater might be the contribution from EDTA added in algal bags (3.4 mg/L) in every 2 days as well as from carbon released by bacteria or external contamination.

Table 6.3: Typical characteristic of the algal culture (algae + AOM), seawater, and UF feed water

Parameters	Units	Algal bags (Algae + AOM)	North Seawater	UF feed (AOM + seawater)
Algal cell concentration	cells/mL	12,000 ± 3,500	600	-
Chlorophyll-a concentration	µg/L	72 ± 38	BDL	BDL
Biopolymer concentration	mg - C/L	1.8 ± 0.2	0.2 ± 0.12	0.5 ± 0.02
Phosphate concentration	µgPO$_4$-P/L			45.2 ± 7.8
MFI-UF$_{10kDa}$*	s/L^2	n.m	7,250	12,610
pH		8.5	8.1	7.8 - 8.0
Temperature	°C	n.m	13 -14	18 - 23

* Modified fouling index (MFI-UF$_{10kDa}$) was measured at a constant filtration flux of 60 L/m^2/h

6.3.2.2 Pressure drop (ΔP) development in MFS cells

The applicability of PRT™ in combination with tight UF (10 kDa) towards reducing the biofouling of SWRO membranes was studied using membrane fouling simulator (MFS) in a pilot plant. Three MFS monitors were run in parallel with different feed solution such as; i) UF feed (MFS 1), ii) permeate of 10 kDa UF (MFS 2), and iii) permeate of tight UF + PRT™ (MFS 3) as shown in Figure 6.9. All three MFS monitors were operated under identical conditions. Within the research period of 21 days, no increase in pressure drop was observed in the MFS 3 monitor fed with permeate of tight UF + PRT™ (Figure 6.9a). However, the pressure drop increase of approximately 500 mbar was observed in MFS 2 monitor fed with permeate of 10 kDa after 17 days of operation (Figure 6.9 b). Likewise, the MFS 1 fed with UF feed water showed an exponential pressure drop increase from the start of the experiment and recorded the increase in pressure drop of about 1700 mbar.

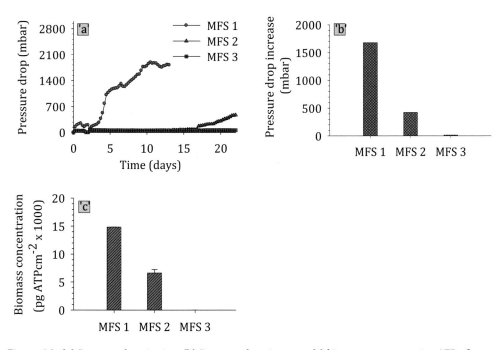

Figure 6.9: (a) Pressure drop in time (b) Pressure drop increase (c) biomass concentration ATP after 21 days of operation fed with UF feed solution (MFS 1), tight UF (10 kDa) permeate (MFS 2) and tight UF (10 kDa) + PRT™ permeate (MFS 3)

The increase in pressure drop after days 17 in the MFS monitor fed with permeate of 10 kDa UF was most likely due to biofouling. It was demonstrated by the result of the membrane autopsy performed at the end of the experiments, which showed the accumulated biomass of about 6,000 pg ATP/cm². The reason for the occurrence of biofouling could be; i) the passage of the low molecular weight organics and ii) the passage of soluble phosphate through ultrafiltration membrane. The hypothesis that the biofouling can be controlled by removing phosphate was verified by observing no pressure drop increase in MFS 3 (Figure 6.9a). The membrane autopsy performed at the end of experiment also showed that the accumulated biomass measured by pg ATP/cm² was below detection limit. The further verification of the hypothesis was performed by bacterial regrowth potential tests as shown in Figure 6.10. As illustrated in Figure 6.10, the application of UF combined with PRT™ (sample B) reduced the net bacterial regrowth from 8 to 0.77 million cells/mL, while the reduction was from only 8 to 1.72 million cells/mL only with the application tight UF (sample A). This suggested the role of phosphate removal by PRT™ towards the reduction potential of the net bacterial regrowth. This was further verified by the observed higher net bacterial growth (0.77 to 1.72 million cells/mL) when the bacterial growth tests were performed in a permeate sample of PRT™ spiked with 10 µg PO₄-P/L.

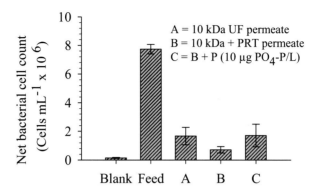

Figure 6.10: (a) Net bacterial regrowth of inoculated natural consortium of marine bacteria in blank (ASW), feed (AOM produced by Chaetoceros affinis, permeate of 10 kDa UF, permeate of 10 kDa UF combined with PRT™, and permeate of 10 kDa UF + PRT™ spiked with 10 µg PO₄-P/L

Furthermore, the increased pressure drop in MFS 2 when fed with permeates of 10 kDa UF could be attributed to the contribution of the passage of low molecular weight (LMW) organics through UF. The LC-OCD analyses of AOM produced by *Chaetoceros affinis,* the feed solution during UF filtration, confirmed the presence of LMW organics in the AOM (Figure 6.11a). The LMW organics might have originated from the EDTA (290 Da), vitamin B12 (1,355 Da), and biotin (244 Da) added to the algal culture. A preliminary investigation was performed to determine the contribution of EDTA on bacterial regrowth potential. To demonstrate this, synthetic seawater prepared was spiked with different EDTA concentration (0 - 10 mg/L) and vitamins. The samples were inoculated with a natural bacterial consortium (10^4 cells/mL), incubated (at 30 °C) and bacterial regrowth was monitored using flow cytometry. The net bacterial regrowth was calculated by subtracting the bacterial cell concentration at day 0 from the maximum bacterial cell concentration during the incubation period. The calculated net bacterial regrowth versus EDTA concentrations was plotted (Figure 6.11b). Interestingly, EDTA concentration correlated moderately with the net bacterial regrowth (R^2 = 0.65), which might be an indication that EDTA contributes to the occurrence of biofouling. Kharusi et al. (2016) also studied an effect of EDTA on the growth of consortium bacteria and found that some strains grew at 0.1 mM of EDTA concentration and some did not grow (Al Kharusi et al., 2016). Likewise, Amon *et al.* (1996) also reported that bacteria could utilize both low molecular weight (LMW < 1 kDa) and high molecular weight (HMW > 1 kDa) organic compounds (Amon et al., 1996). Moreover, detail study to foresee the contribution of EDTA on biofouling is essential.

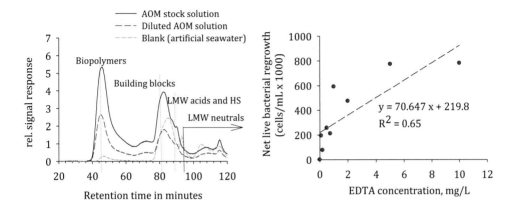

Figure 6.11: a) LC-OCD chromatograms of AOM extracted from marine bloom-forming algae species Chaetoceros affinis (2 x diluted with ASW and non-diluted), and blank (ASW), b) Relationship between the net bacterial regrowth and EDTA concentration

6.3.2.3 Assessment of permeate quality of UF-PRT™

The feed and permeate samples were collected once in every week from the demonstration UF-PRT™ pilot plant for 21 days. The collected samples were analyzed in term of biopolymer removal, phosphate removal, and reduction in bacterial regrowth potential (BRP). As illustrated in Figure 6.12a, the biopolymer rejection by tight UF (10 kDa) was approximately 95 %, which was similar to the result obtained in the laboratory scale experiments (refer to Figure 6.5a). The biopolymer rejection was slightly higher when tight UF was combined with PRT™ (Table 6.4). Interestingly, the standard deviation of the measured phosphate was slightly higher, which showed the variation of phosphate during the experimental period. The reason for the variation might be explained by the passage of some algal cells from the algal bag to the UF feed tank. These algal cells might have uptake the phosphate available in the UF tank for its growth in the tank. It has been reported that algae can remove the phosphate due to adsorption and algae-induced chemical precipitation (Sanudo-Wilhelmy et al., 2004). It was also observed that after 10 days of operation there was the algal growth in the UF feed tank of UF-PRT™ and thus the UF operation was stopped to clean the UF feed tank. Moreover, the operation of PRT™ was continued using the buffer capacity of the UF permeate tank.

Likewise, the measured phosphate concentration in permeate of PRT™ was approximately 4 µg PO₄-P/L, which showed the removal efficiency of > 95 % (Figure 6.12b). The measure absolute phosphate concentration of 4 µg PO₄-P/L in permeate of PRT™ was similar to the detection limit of the Skalar San++ analyzer (refer section 6.3.4).

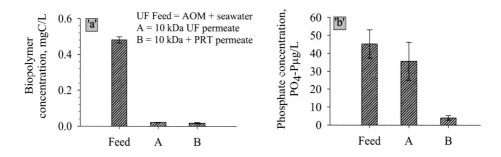

Figure 6.12: (a) Biopolymer concentration and (b) Phosphate concentration measured in feed (AOM + seawater), 10 kDa UF permeate, and in permeate of 10 kDa UF+PRT^{TM}

The summary of the measured parameters is presented in Table 6.4.

Table 6.4: Summary of biopolymer concentration and phosphate concentration and percentage of reduction by 10 kDa UF and PRT^{TM}

Sample descriptions	Biopolymer concentration		Phosphate concentration	
	(mg - C/L)	Removal (%)	in PO₄-P (µg/L)	Removal (%)
Seawater	0.2 ± 0.12		24.9 ± 11.50	
AOM	1.8 ± 0.2		417 ± 150	
UF feed (Seawater + AOM)	0.5 ± 0.02		94.0 ± 27.80	
10 kDa UF permeate	0.02	95.6	44.2 ± 10.60	47.0
10 kDa UF + PRT™ permeate	0.02	96.4	4.0 ± 1.30	95.7

This "pilot plant investigation" also clearly shows that the removal of phosphate by the application of PRT™ combined with UF (10 kDa) restrict biomass growth and thus delay/eliminate the occurrence of biofouling in SWRO membranes. The tested resin in this study was the bio-based weakly anion exchange resin coated with iron hydroxide group. The XRD analysis of the coated surface of the resin showed the presence of crystalline goethite. Furthermore, the inductively coupled plasma mass spectrometry (ICP-MS) analysis performed also indicated that the coated surface of the resins consists of 97.4 % of Fe. The work performed by Tejedor-Tejedor et al. (1990) and Belelli et al. (2014) had reported that the phosphates are known to form strongly bonded complexes with goethite surface in the wide range of pH (Belelli et al., 2014, Tejedor-Tejedor et al., 1990). The presence of the functional group in the coated surface of the resin was investigated by performing the FTIR analysis. As presented in Figure 6.13, the FTIR spectra of resin showed the presence of high absorption broad and intense band (peak A) corresponding to the stretching O-H at a wavelength of 3,336 cm⁻¹. Other bands that appeared in the wavelength ranged from 1,083 to 2,926 cm⁻¹ corresponds to lipids, proteins, and humic substances (Mecozzi et al., 2001, Villacorte et al., 2015a). The broad and intense band of O-H forms presents on the surface goethite (resins) forms a surface complex with phosphate.

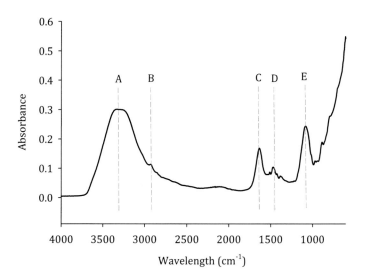

Figure 6.13: FTIR spectra of PRT^(TM) adsorbent

The potential of bio-based phosphate adsorbent towards the removal of phosphate demonstrated the possible application for controlling biofouling in SWRO systems. The theoretical calculation showed that breakthrough of adsorbent would occur after 99 days of operation. The calculation was based on the assumption that the adsorption capacity of the resin is 1.75 mg PO_4/g, as provided by the supplier.

Feed flow rate	= 9.12 L/h (see section 6.2.3.2)
Bulk density of PRT™ media	= 0.765 g/mL
Mass = Bed volume x bulk density	= 1,163.21 g
[PO_4] feed	= 0.094 mgPO_4/L
PO_4 capacity = mass of media x adsorption capacity	= 2,035.6 mgPO_4
Volume treated = PO_4 capacity / [PO_4] feed	= 21,655.4 L
Estimated time for breakthrough of media	= Volume treated /feed flow rate
	= 99 days

The bio-based phosphate adsorbent is fully regenerable. The regeneration can be performed with 2 % by weight of NaOH solution according to the following protocol suggested by the adsorbent supplier;

Backwash: The system is the first backwash with demi-water/tap water for 15-30 minutes at a flow rate of 10 L/h to unpack the bed and break the clusters of resin. In this process, the suspended solids trapped in the upper part of the bed during the adsorption cycle are also removed.

Countercurrent desorption: The column is regenerated in a counter-current mode with 2 % w NaOH for 3 hours at a low flow rate of 4 L/h. The bed is allowed for 20 % to allow a better mixing in the column and to disaggregate the resin.

Rinse: Demi-water/tap water is flushed through the resin bed in up-flow mode whereby the resin is fluidized, and resin bed is expanded. The bed is rinsed until the pH drops to the desired value, i.e., approximately for 5 hours at a flow rate of 10 L/h.

6.4 Conclusions

Evaluation of PRT™ at laboratory scale and pilot scale led to the following conclusions;

- Application of PRT™ removed approximately 98 % of the dissolved phosphate from feed water. The measured phosphate in permeate of PRT™ was approximately 4 - 5 µg PO₄ - P/L, which was the detection limit of the method applied. The measured lower phosphate in PRT™ coincided with the lower bacterial regrowth potential, which is an indication that phosphate limits the bacterial regrowth. It was verified by the experiment that when permeate of tight UF+PRT™ was spiked with 10 µg PO₄ - P/L, the bacterial regrowth increased from 0.7 to 1.7 million cells/L.

• Biofouling experiment performed with membrane fouling simulator (MFS) showed no increase in feed channel pressure drop in MFS fed with permeate of tight UF +PRT™ for at least a period of 21 days. While the MFS operated with permeate of tight UF alone showed an increase in pressure drop of approximately 500 mbar after 17 days of operation. This demonstrated that the application of tight UF +PRT™ has the potential in delaying the onset of organic/biological fouling in SWRO feed water during algal blooms.

• The increase in pressure drop in MFS cell fed with permeate of tight UF alone may be due to the presence of low molecular weight (LMW) organic as well as dissolved phosphate. A preliminary investigation performed with a various concentration of EDTA (origin of LMW) showed a moderate linear correlation (R^2 = 0.65) with the measured net bacterial regrowth potential.

• The measured bacterial regrowth potential (BRP) in permeate of UF + PRT™ was found independent of the pore size of the UF combined with PRT™.

6.5   Acknowledgements

Thanks to Marco Dubbeldam and Bernd van Broekhoven from Zeeschelp for algal culture, Tom Spanier, Henry Hamberg, Sander Brinks from Pentair X-flow and Sandie Chauveau from BiAqua for their technical support in a pilot plant, Almotasembellah Abushaban for ATP measurement.

6.6   Reference

Al Kharusi, S., Abed, R.M.M. and Dobretsov, S. (2016) EDTA addition enhances bacterial respiration activities and hydrocarbon degradation in bioaugmented and non-bioaugmented oil-contaminated desert soils. Chemosphere 147(Supplement C), 279-286.

Amon, R.M.W. and Benner, R. (1996) Bacterial utilization of different size classes of dissolved organic matter. Limnology and Oceanography 41(1), 41-51.

Baker, J.S. and Dudley, L.Y. (1998) Conference Membranes in Drinking and Industrial Water ProductionBiofouling in membrane systems — A review. Desalination 118(1), 81-89.

Belelli, P.G., Fuente, S.A. and Castellani, N.J. (2014) Phosphate adsorption on goethite and Al-rich goethite. Computational Materials Science 85, 59-66.

Dreszer, C., Flemming, H.C., Zwijnenburg, A., Kruithof, J.C. and Vrouwenvelder, J.S. (2014) Impact of biofilm accumulation on transmembrane and feed channel pressure drop: Effects of crossflow velocity, feed spacer and biodegradable nutrient. Water Research 50, 200-211.

Flemming, H.C. (2011) Biofilm highlights, Springer - Verlag Berlin Heidelberg.

Flemming, H.C. and Schaule, G. (1988) Biofouling on membranes - A microbiological approach. Desalination 70(1), 95-119.

Flemming, H.C., Schaule, G., Griebe, T., Schmitt, J. and Tamachkiarowa, A. (1997) Biofouling—the Achilles heel of membrane processes. Desalination 113(2–3), 215-225.

Holtan, H., Kamp-Nielsen, L. and Stuanes, A.O. (1988) Phosphorus in soil, water and sediment: An overview. Hydrobiologia 170, 19-34.

Huber, S.A., Balz, A., Abert, M. and Pronk, W. (2011) Characterisation of aquatic humic and non-humic matter with size-exclusion chromatography – organic carbon detection – organic nitrogen detection (LC-OCD-OND). Water Research 45(2), 879-885.

Jacobson, J.D., Kennedy, M.D., Amy, G. and Schippers, J.C. (2009) Phosphate limitation in reverse osmosis: An option to control biofouling? Desalination and Water Treatment 5, 198-206.

Kooij, D.V.d., Visser, A. and Hijnen, W.A.M. (1982) Determination of easily assimilable organic carbon in drinking water. Journal of the American Water Works Association. 74, 540-545.

Maher, W. and Woo, L. (1998) Procedures for the storage and digestion of natural waters for the determination of filterable reactive phosphorous, total fiterable phosphorous and total phosphorous. Analytica Chimica Acta 375, 5-47.

Matin, A., Khan, Z., Zaidi, S.M.J. and Boyce, M.C. (2011) Biofouling in reverse osmosis membranes for seawater desalination: Phenomena and prevention. Desalination 281, 1-16.

Mecozzi, M., Acquistucci, R., Di Noto, V., Pietrantonio, E., Amici, M. and Cardarilli, D. (2001) Characterization of mucilage aggregates in Adriatic and Tyrrhenian Sea: structure similarities between mucilage samples and the insoluble fractions of marine humic substance. Chemosphere 44(4), 709-720.

Nguyen, T., Roddick, F.A. and Fan, L. (2012) Biofouling of Water Treatment Membranes: A Review of the Underlying Causes, Monitoring Techniques and Control Measures. Membranes 2, 804-840.

Radu, A.I., Vrouwenvelder, J.S., van Loosdrecht, M.C.M. and Picioreanu, C. (2012) Effect of flow velocity, substrate concentration and hydraulic cleaning on biofouling of reverse osmosis feed channels. Chemical Engineering Journal 188, 30-39.

Sanudo-Wilhelmy, S.A., Tovar-Sanchez, A., Fu, F.-X., Capone, D.G., Carpenter, E.J. and Hutchins, D.A. (2004) The impact of surface-adsorbed phosphorus on phytoplankton Redfield stoichiometry. Nature 432(7019), 897-901.

Sevcenco, A.-M., Paravidino, M., Vrouwenvelder, J.S., Wolterbeek, H.T., van Loosdrecht, M.C.M. and Hagen, W.R. (2015) Phosphate and arsenate removal efficiency by thermostable ferritin enzyme from Pyrococcus furiosus using radioisotopes. Water Research 76, 181-186.

Tejedor-Tejedor, M.I. and Anderson, M.A. (1990) The protonation of phosphate on the surface of goethite as studied by CIR-FTIR and electrophoretic mobility. Langmuir 6(3), 602-611.

Ugurlu, A. and Salman, B. (1998) Phosphorus removal by fly ash. Environment International 24(8), 911-918.

van Loosdrecht, M.C.M., Bereschenko, L., Radu, A., Kruithof, J.C., Picioreanu, C., Johns, M.L. and Vrouwenvelder, H.S. (2012) New approaches to characterizing and understanding biofouling of spiral wound membrane systems. Water Science and Technology 66(1), 88-94.

Villacorte, L.O. (2014) Algal blooms and membrane based desalination technology, Ph.D thesis, UNESCO-IHE/TUDelft, Delft

Villacorte, L.O., Ekowati, Y., Neu, T.R., Kleijn, J.M., Winters, H., Amy, G., Schippers, J.C. and Kennedy, M.D. (2015a) Characterisation of algal organic matter produced by bloom-forming marine and freshwater algae. Water Research 73, 216-230.

Villacorte, L.O., Tabatabai, S.A.A., Dhakal, N., Amy, G., Schippers, J.C. and Kennedy, M.D. (2015b) Algal blooms: an emerging threat to seawater reverse osmosis desalination. Desalination and Water Treatment 55(10), 2601-2611.

Vrouwenvelder, J.S., Beyer, F., Dahmani, K., Hasan, N., Galjaard, G., Kruithof, J.C. and Van Loosdrecht, M.C.M. (2010) Phosphate limitation to control biofouling. Water Research 44(11), 3454-3466.

Vrouwenvelder, J.S., Graf von der Schulenburg, D.A., Kruithof, J.C., Johns, M.L. and van Loosdrecht, M.C.M. (2009) Biofouling of spiral-wound nanofiltration and reverse osmosis membranes: A feed spacer problem. Water Research 43(3), 583-594.

Vrouwenvelder, J.S., van Paassen, J.A.M., Wessels, L.P., van Dam, A.F. and Bakker, S.M. (2006) The Membrane Fouling Simulator: A practical tool for fouling prediction and control. Journal of Membrane Science 281(1–2), 316-324.

Annexes: Supporting information

Challenges during the operation of pilot plants

The pilot plant operation had several challenges during the Ph.D. period

i. Frequent shut down of the ultrafiltration operation

The operation of UF was frequently stopped which resulted in no permeate production. This had consequences for the operation of downstream Membrane Fouling Simulator (MFS). The capacity of UF permeates tank was 20 L, which was enough for 2.5 hours operation of MFS at a cross flow velocity of 0.2 m/s considering the dead volume needed for a startup the UF operation.

ii. Leaking from Amiad strainer

During the pilot plant operation, Amiad strainer had a leakage problem. The Amiad filter was also damaged many times mainly by the deposition of silt passing through the intake pump.

iii. Location of pilot

Physical access to the pilot plant located in Jacobahaven, the Netherlands was an issue especially for those who do not have a driving license.

iv. Algal culture

The continuous production of algal organic matter needed for a long-term operation of UF and MFS was a challenge during the pilot operation in both time and cost.

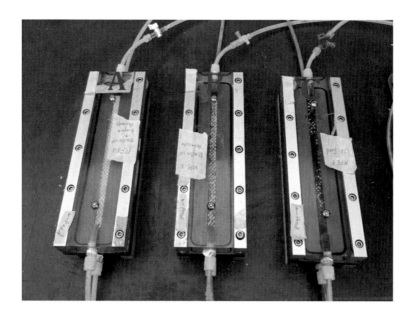

Supplementary Figure S.6.1: Three MFS cells on operation fed with A) 10 kDa UF+PRT™ permeate, B) 10 kDa UF permeate, and C) UF feed

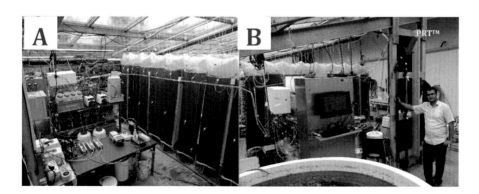

Supplementary Figure S.6.2: A) General overview of a pilot plant in Zeeland, Jacobahaven, the Netherlands. The large plastic bags are the algal culture of marine diatom chaetoceros affinis B) PRT™ column packed with phosphate adsorbent

7

Conclusions and outlook

Contents

7.1   Conclusions

Seawater desalination using reverse osmosis technology is currently dominating the desalination market and is widely applied for both drinking water and industrial water production. The total desalination capacity (installed and projected for 2018) is about 80 million m^3/day, of which 75 % ($\sim$ 60 million m^3/day) uses reverse osmosis technology. In the near future, the ratio will likely change in favor of reverse osmosis since most new contracted desalination plants use membrane-based technology (Desal Data, 2016).

An emerging threat to SWRO operation is membrane fouling caused by algal blooms (a so-called "population explosion" of algae). During an algal bloom, algae release algal organic matter (AOM) that can cause membrane fouling. Consequently, SWRO plants are equipped with pre-treatment systems (e.g., media filters with/without coagulation, dissolved air flotation with coagulation and filtration and/or ultrafiltration with/without coagulation) to lower the fouling potential of SWRO feed water. Fouling resulting from (inorganic) particulate matter is well controlled by conventional pre-treatment systems. However, the occurrence of organic and biological fouling is still a major issue in SWRO membrane systems mainly due to poor removal efficiency of AOM and dissolved nutrients such as carbon and phosphate in pretreatment systems.

The failure of pre-treatment to provide sufficient and acceptable feed water quality for SWRO during algal blooms has gained the attention of the desalination industry. It is expected that pre-treatment systems that can completely remove AOM, as well as nutrients such as dissolved phosphate and carbon from SWRO feed water, can delay the *onset* of organic and biofouling in SWRO systems. Furthermore, better *methods/tools* need to be developed to assess and improve pre-treatment processes in terms of their ability to reduce re-growth potential prior to SWRO membranes.

In this study, the following approaches were taken:

- A method to measure bacterial regrowth potential (BRP) in seawater, which makes use of a natural bacterial consortium as inoculum in combination with flow cytometry, was further developed. Improvements to the method were required to lower the limit of detection and improve sensitivity.

- Understanding ultrafiltration membrane fouling and the root causes of poor backwashability resulting from organic matter generated by four different marine algal species.

- Assessing the ability of conventional UF (150 kDa) and tight UF (10 kDa) alone and in combination with phosphate removal technology (PRT™) in reducing bacterial regrowth potential and delaying the onset of organic/biological fouling in SWRO feed water during algal blooms

7.1.1. Developing an improved Bacterial Regrowth Potential (BRP) method

An improved method to measure bacterial regrowth potential (BRP) in seawater was developed. The method is based on flow cytometry combined with fluorescence staining (SYBR® Green I and Propidium Iodide) and makes use of a natural bacterial consortium as inoculum. The use of a natural bacterial consortium provides a broader and diverse substrate range compared with a single pure culture, potentially offering a more realistic interpretation of biofouling occurring in SWRO membranes. The BRP method studied is based on the method used by (Dixon et al., 2012) for seawater and (Prest et al., 2013) for freshwater. The BRP method was originally developed by (Withers et al., 1998; Sathasivan et al., 1999, for freshwater. The method involves the removal/inactivation of bacteria (live + dead) from the water sample, followed by re-inoculation with a natural consortium of marine bacteria (10^4 cells/mL), with incubation (30 °C), and monitoring of microbial regrowth using flow cytometry. The net live bacterial regrowth calculated from the bacterial growth curve was considered as an indicator of bacterial regrowth potential.

The level of detection (LOD) of the method was lowered by developing a standard protocol to prepare blank seawater. The following aspects were considered:

- Minimize the level of contamination that might originate from sample bottles, chemicals, pipettes and the laboratory environment during blank seawater preparation.
- Minimize leaching of carbon from filters and all surfaces during BRP measurements.

The limit of detection of the improved BRP method was lowered to $43 \times 10^3 \pm 12 \times 10^3$ cells/mL. The method was validated by performing a calibration of bacterial regrowth potential with glucose as a standard substrate in artificial and natural seawater. The calibration of BRP showed good linearity (R^2 = 0.88 to 0.95) for a range of glucose concentration (0 – 2,000 μg-$C_{glucose}$/L). The calibration curve was used to calculate the yield factor of the marine bacteria and converted to an equivalent carbon concentration. Accordingly, the lowest measured equivalent carbon concentration with this method was 9.3 ± 2.6 μg-$C_{glucose}$/L. Furthermore, the developed method was applied in full-scale seawater desalination plants to assess the efficiency of pre-treatment systems in reducing bacterial regrowth potential of SWRO feed water. Results showed a bacterial regrowth potential reduction of 54 % in SWRO feed water with DAF-UF as pre-treatment and 40 % reduction with DMF-CF. Furthermore, the bacterial regrowth potential values of SWRO feed water after DAF-UF-pretreatment coincided with a frequency of chemical cleaning of SWRO membranes. However, much more research is needed to investigate if a clear correlation exists between bacterial regrowth potential and the rate of biofouling development in SWRO systems.

7.1.2. Fouling potential of marine bloom-forming algae species

The fouling potential of four marine algal species *(Chaetoceros affinis, Rhodomonas balthica, Tetraselmis suecica,* and *Phaeocystis globulosa*) at different phases of their growth was investigated. Parameters measured during the investigations were algal cell density, chlorophyll-a concentration, biopolymer concentration (LC-OCD), transparent exopolymer particles (TEP$_{10kDa}$), modified fouling index (MFI-UF$_{10kDa}$), bacterial re-growth potential, etc.

Batch culture monitoring of all four algal species illustrated varying growth patterns, algal cell density, transparent exopolymer particles (TEP$_{10kDa}$), biopolymer concentration, and modified fouling index (MFI-UF$_{10kDa}$). Among the four algal species, *Chaetoceros affinis* showed the highest biopolymer concentration, TEP$_{10kDa}$, and MFI-UF$_{10kDa}$. Results showed that the membrane fouling potential (MFI-UF) correlated (R^2 > 0.95) with TEP, biopolymer concentration, algal cell density and chlorophyll-a concentration during the growth phase of algal species. However, no apparent relationship of MFI-UF with algal cell density and chlorophyll-a concentration was observed during the death phase. This indicates that future management practices of membrane-based desalination plants should include the monitoring of the following parameters: TEP$_{10kDa}$, MFI-UF$_{10kDa}$, and biopolymer concentration to better predict plant performance and developed preventive and corrective membrane fouling strategies.

7.1.3. UF fouling by algal released organic matter (AOM)

Ultrafiltration membrane fouling caused by algal organic matter (AOM) released by four marine bloom-forming algae species *(Chaetoceros affinis, Rhodomonas balthica, Tetraselmis suecica,* and *Phaeocystis globulosa)* was also investigated. The AOM containing 0.5 mg-biopolymer-C/L was fed to hollow fiber polyethersulphone UF membranes (150 kDa). The operation was dead-end filtration and inside-outside mode at a constant flux of 80 L/m^2/h. Result illustrated that the AOM solution with higher TEP concentration resulted in higher non-backwashable fouling in UF (150 kDa) while operating without coagulation. The non-backwashable fouling was mainly caused by polysaccharides (OH), and sugar ester group presents in the AOM as illustrated by FTIR results of fouled and (physically) cleaned membranes. In addition, the MFI-UF$_{150kDa}$ calculated from the slope of first UF filtration cycle also correlated with the non-backwashable fouling rate in UF. Moreover, results also demonstrated that non-backwashable fouling in UF depends on the type of bloom-forming algae and the nature of AOM they release. For instance, the rate of non-backwashable fouling with AOM of *Rhodomonas balthica* was 0.07 bar/h, *Chaetoceros affinis* was 0.05 bar/h, *Tetraselmis suecica* was 0.03 bar/h, and *Phaeocystis globulosa* was 0.01 bar/h. Finally, the analysis of permeate quality of standard ultrafiltration membrane when operated without coagulation illustrated the passage of 20 – 40 % of the biopolymer. The consequences of which might be the potential of organic and biofouling occurrence in downstream SWRO membranes. This suggests the need for more robust pre-treatment that can better remove biopolymer fractions as a future pre-treatment for SWRO membranes.

7.1.4. Role of tight UF (10 kDa) as a pre-treatment during algal bloom

The application of tight ultrafiltration (10 kDa) as an alternative pre-treatment during algal blooms was investigated and compared with various pore size MF and UF membranes. The proof of principle was performed at laboratory and pilot scale using flat sheet membranes (MF: 0.1 µm (PVDF), UF: 100, 30, and 10 kDa (RC) and hollow fiber membranes (UF: 10

and 150 kDa PES). The feed solution used was algal organic matter (AOM) produced by *Chaetoceros affinis*. In all cases, the feed and permeate water quality were assessed in terms of biopolymer concentration, TEP_{10kDa}, $MFI\text{-}UF_{10kDa}$, and bacterial regrowth potential.

Results illustrated that the rejection of algal biopolymer produced by *Chaetocerous affinis* was 3 to 4 times higher with tight UF (10 kDa) compared to the higher molecular weight cut off MF/UF membranes. The biopolymer fraction of AOM showed a linear correlation ($R^2 = 0.88$) with net bacterial regrowth. The LC-OCD analysis of the sample at day 0 and day 27 of incubation during the bacterial regrowth potential test showed that arger biopolymers were more easily degraded than smaller ones. On the contrary, no significant difference was observed between the net bacterial regrowth in permeate of tight UF (10 kDa) compared with conventional UF (150 kDa) during testing at pilot scale.

Furthermore, biofouling experiments were performed using membrane fouling simulators (MFS). Three MFS cells were fed with *i*) UF feed without pre-treatment, *ii*) permeate of conventional hollow fiber UF (150 kDa), and *iii*) permeate of tight UF (10 kDa). No substantial increase in feed channel pressure drop was observed over a period of 15 days when 150 kDa and 10 kDa UF permeates were fed to MFS monitors at a cross flow velocity of 0.2 m/s. Moreover, the biomass accumulated in the MFS fed with permeate of tight UF (10 kDa) was about 860 pg ATP/cm^2, which was *ca.* 2 times lower than in the MFS fed with permeate of a 150 kDa UF. Overall, the results illustrate the potential of tight UF to delay the onset of biofouling in SWRO membranes.

Nevertheless, the non-backwashable fouling rates development after each succeeding CEB cycles were approximately 1.5 times higher in the tight UF (10 kDa) compared to the conventional UF (150 kDa). Therefore, it is still important to improve the backwashability of the tight UF membrane for future applications. It is expected that improving the surface porosity of the membrane can improve the backwashability and allow better removal of the cake/gel layer formed on the membrane surface during UF.

7.1.5. Phosphate removal and bacterial regrowth potential of SWRO feed water

The application of an adsorbent to remove phosphate (PRT™) combined with tight UF (10 kDa) to reduce bacterial regrowth potential was investigated at laboratory and pilot scale.

Algal bloom conditions were simulated by culturing the marine diatom *Chaetoceros affinis*. Experiments were performed with algal organic matter (AOM) harvested from bloom-forming marine algae (*Chaetoceros affinis*).

Results demonstrated that the application of a phosphate adsorbent (PRT™) reduced the phosphate level to 4 - 5 µgPO$_4$-P/L, which is approximately 98 % removal in column experiments designed at an EBCT of 10 minutes. The final phosphate level in permeate is still much higher than the recommended threshold value below which biofouling would not occur. Moreover, 5 µgPO$_4$-P/L is the current detection limit of the Skalar San^{++} analyzer used in this research. Consequently, there is a need to further lower the level of detection of phosphate in seawater. Moreover, it was also observed that the lower the phosphate level after the phosphate adsorbent (PRT™), the lower the bacterial regrowth potential observed. The addition of 10 µgPO$_4$-P/L to PRT™ permeate increased the bacterial regrowth by 55 %, suggesting that limiting phosphate may delay the onset of biofouling. To verify this, accelerated (21 days) biofouling experiments, simulating similar conditions to those in SWRO systems, were performed using membrane-fouling simulator (MFS). The experimental investigation showed that removal of nutrients such as carbon by tight UF (10 kDa) and phosphate by PRT™ lowered fouling rate development. Biofouling experiments performed in MFS monitors showed no increase in feed channel pressure drop over a period of 21 days when fed with the permeate of tight UF followed by PRT™ at a cross flow velocity of 0.2 m/s.

On the other hand, the MFS operated with permeate of tight UF alone showed an increase in pressure drop of approximately 500 mbar after 17 days of operation. The increase in pressure drop in MFS cells fed with the permeate of tight UF could possibly be due to the passage of low molecular weight (LMW) organics (used to grow algae) as well as dissolved phosphate. A preliminary investigation performed with a range of EDTA concentrations (origin of LMW) showed a moderate linear correlation (R^2 = 0.65) with the measured net bacterial regrowth potential. This suggests that the application of PRT™ downstream of tight UF could be a good option to delay the onset of biofouling in SWRO systems. Moreover, long-term experiments are required to verify these preliminary findings.

7.2   General outlook

This study demonstrated that an improved bacterial regrowth potential (BRP) method can be used to (*i*) assess pre-treatment technology in terms of BRP reduction, (*ii*) monitor the performance of pre-treatment systems, and (*iii*) develop essential strategies to mitigate membrane fouling in SWRO systems. The study highlighted that the future improved best management practices of membrane-based desalination plants should include regularly monitoring of TEP, MFI-UF, biopolymer concentration, and BRP in raw water so that corrective membrane fouling strategies can be developed in time. However, determining MFI-UF, biopolymer, TEP concentration, BRP during algal blooms and correlating with the MODIS satellite data might generate useful information to predict the fouling potential of seawater remotely at different locations in the future.

This study also demonstrated that the removal of algal organic matter (AOM), and dissolved phosphate from SWRO feed water is a potential strategy to delay the onset of organic and biofouling in SWRO systems during algal blooms. Tight UF (10 kDa) coupled with an adsorbent to remove phosphate showed higher potential compared to UF alone (10 kDa) with respect to AOM and nutrient (C, P) removal.

Furthermore, research on the long-term operational stability of SWRO plants during algal blooms operating at different fluxes, coagulant dosing, etc is still needed. Further improvement in UF material properties should also be considered to minimize the cost of operation and ensure stable hydraulic operation. Finally, it is still necessary to further develop existing and new methods that can detect low concentrations of nutrients e.g. carbon and phosphate in seawater, to support the development of improved membrane fouling prevention strategies.

7.3   References

Dixon, M.B., Qiu, T., Blaikie, M. and Pelekani, C. (2012) The application of the Bacterial Regrowth Potential method and Flow Cytometry for biofouling detection at the Penneshaw Desalination Plant in South Australia. Desalination 284, 245-252.

Prest, E.I., Hammes, F., Kötzsch, S., van Loosdrecht, M.C.M. and Vrouwenvelder, J.S. (2013) Monitoring microbiological changes in drinking water systems using a fast and reproducible flow cytometric method. Water Research 47(19), 7131-7142.

Sathasivan, A. and Ohgaki, S. (1999) Application of new bacterial regrowth potential method for water distribution system – a clear evidence of phosphorus limitation. Water Research 33(1), 137-144.

Withers, N. and Werner, P. (1998) Bacterial regrowth potential:quantitative measure by acetate carbon equivalents. Water 25 (5), 19-23.

Samenvatting

Seawater reverse osmosis (SWRO) is tegenwoordig de meest gebruikte technologie voor zeewater desalination. Toch blijft membrane fouling een flinke uitdaging voor de rendabele werking van SWRO. Een veel voorkomend obstakel bij SWRO membrane fouling (organische/biofouling) is het optreden van algal bloom, die de concentratie van algal cells en algal organic matter (AOM) in zeewater laten toenemen.

Om de membrane fouling te minimaliseren zijn SWRO systemen voorzien van pre-treatment systemen. Desondanks zijn de huidige pre-treatment systemen niet in staat om alle AOM uit SWRO feed water te verwijderen. De AOM dat van de pre-treatment systemen afkomt, verzamelt zich op de SWRO membrane oppervlaktes en werkt als een ''conditioning layer'' om de biofilm ontwikkeling, met de beschikbare voedingstoffen (C, P) in het RO feed water, op gang te brengen. Een belangrijk voorbeeld hiervan is de red tide algal bloom in de Golf staat in het Midden-Oosten tussen 2008 - 2009. In deze periode raakten verschillende pre-treatments, zoals granular media filter (GMF) met coagulatie, verstopt. Tegelijkertijd produceerde de pre-treatments een slechte kwaliteit aan SWRO feed water (SDI > 5). Als gevolg hiervan was het van belang dat bepaalde SWRO desalination plants in de kuststreken gesloten werden om de onomkeerbare fouling van RO membranes te kunnen vermijden. Hierna werd de toepassing van low-pressure membranes, zoals microfiltratie en ultra-filtratie (MF/UF), beschouwd als de meest betrouwbare pre-treatment gedurende algal bloom. Eerder werd al bewezen dat conventionele UF membranes niet geschikt zijn voor het verwijderen van alle organic matter (AOM) van SWRO feed water, en, daarmee, kan organic/biofouling in downstream SWRO optreden. Dit geeft aan dat nieuwe robust pre-treatment technologien, die AOM en andere voedingstoffen (C, P) van SWRO feed water kunnen verwijderen, van belang zijn voor het vertragen of zelfs het tegengaan van organic/biofouling in SWRO systemen.

Dit onderzoek werd voornamelijk gericht op de toepassing van vaste UF (met een moleculair gewicht afgerond op (10 kDa), op zichzelf werkend of in combinatie met phosphate verwijderingstechnologie (PRT™) wat dient als een verbeterde voorbehandeling voor het verwijderen of vertragen van organische/biologische vervuiling in SWRO systemen. De hypotheses die hierbij getest werden waren: i) dichte ultrafiltratie (10 kDa) is effectiever in het verwijderen van biologisch afbreekbare organische materie en daarmee de biofouling potentie van SWRO voedingswater verminderen, en ii) de toepassing van PRT™ kan zorgen voor een vermindering van fosfaat in SWRO voedingswater wat uiteindelijk de bacteriële groeipotentie van SWRO voedingswater kan limiteren. Daarbij richt dit onderzoek zich ook op de ontwikkeling van een methode die gebruikt kan worden voor de meting van bacterial regrowth potential (BRP) in zeewater samples, om de biofouling potentie van SWRO voedingswater te beoordelen. De studies gebaseerd op vaste UF werden aangeleverd door Pentair X-flow Nederland, en phosphate verwijderingstechnologie, die getest werd tijdens dit onderzoek, is ontwikkeld door BiAqua Nederland. De PRT™ gebruikte een fosfaat absorbeermiddel bestaande uit een zwakke anionuitwisseling, gecoat met een ijzer hydroxide groep.

Het onderzoek stond allereerst in het teken van de ontwikkeling en/of verbetering van methodes die gebruikt kunnen worden om bacterial regrowth potential (BRP) te meten in zeewater samples. Tijdens de ontwikkeling van deze methodes werd flow cytometry, gecombineerd met fluorescence staining (SYBR Green I and Propidium Iodide) gebruikt met een natuurlijk consortium bestaande uit marine bacteriën als inoculum. De resulterende methode is relatief snel (2 - 3 dagen) vergeleken met conventionele bioassays (12 - 14 dagen). De kalibratie van de methode werd uitgevoerd met glucose als een vaststaand substraat in kunstmatig- en natuurlijk zeewater. De nieuwe methode toegepast in bestaande zeewater desalination machinerie, gevestigd in het Midden-Oosten. Op deze manier kon de biofouling potential van SWRO, en de uitvoering van de pre-treatment systemen, getest worden.

De tweede fase van het onderzoek werd gericht op het interpreteren van de fouling potentie en het fouling gedrag van alg en, in algen-vrijgelaten organische materie, in ultrafiltratiemembranen. Hiervoor werden vier marine algen soorten gebruikt: *Chaetoceros affinis (Ch), Rhodomonas balthica (Rb), Tetraselmis suecica (Te),* en *Phaeocystis globulosa (Ph).* Tijdens de groei en stilstaande/afnemende fase werden de dichtheid van de algen cel, chlorophyll-a, biopolymer, transparent exopolymer particles (TEP) concentratie en MFI-UF$_{10kDa}$ (membrane fouling potential) gemeten. Fouling experimenten werden uitgevoerd met capillaire ultrafiltratie, filtratie van binnen en van buitenaf. Daarnaast werd de backwashable- en non-backwashable fouling gecontroleerd.

Gedurende de groei en stilstaande/afnemende fase van de algensoorten, werden opvallende verschillen geobserveerd in de productie van biopolymers, TEP and MFI-UF$_{10kDa}$. De membraan fouling potentie (MFI-UF$_{10kDa}$) was gelijk gerelateerd aan algencel dichtheid, en, gedurende de groeifase van de algensoorten, ook gerelateerd aan chlorophyll-a concentration, biopolymer concentration, TEP. Na de groeifase werd het duidelijk dat de relatie tussen MFI-UF$_{10kDa}$, algencel dichtheid en chlorophyll-a concentration zich niet verder voortzette. Experimenten waarbij capillaire ultrafiltratie gebruikt werd toonden aan dat membranen (150 kDa) gevoed met water bestaande uit 0.5 mg–biopolymer-C/L back washable fouling, overeenkwamen met de MFI-UF$_{150kDa}$ en TEP voor *Rh, Te, en Ph.* Back washable fouling of Ch week af en was aanzienlijk hoger. De non-back washable fouling van de membranen met ultrafiltratie varieerde hevig met het type algensoort en kwam overeen met MFI-UF$_{150kDa}$ en TEP concentratie. Rh toonde de hoogste en *Ph* de laaste non-backwashable fouling aan (op een level van 0.5 mg-biopolymer-C/L in het voedingswater). Deze non-backwashable fouling toegeschreven aan polysaccharides (uitstrekkend – OH) en suiker ester groep (uitstrekkend S = O) die beide in de AOM aanwezig zijn. Bovendien toonde typering van de permeate kwaliteit van UF de rejectie van biopolymer bij 60 % tot 80 % aan, afhankelijk van de algensoort. Dit laat zien dat de kleinere maat van de biopolymers, in vergelijking tot de maat van de membraanporiën van de ultrafiltratie, ook zou kunnen bijdragen aan de non-backwashable fouling van UF/RO systemen. Hierdoor is er een robuustere voorbehandeling nodig die een hogere concentratie van AOM uit RO voedingswater kan verwijderen en daarmee de organische/biologische fouling in SWRO systemen kan vertragen of zelfs volledig kan tegengaan.

In de daaropvolgende fase van het onderzoek werd de BRP methode en andere analytische middelen, zoals transparante exopolymer particles (TEP), modified fouling index (MFI-UF), liquid chromatography organic carbon detection (LC-OCD), toegepast om de afname van de biofouling potentie van vaste ultrafiltratie (10 kDa) membraan als voorbehandeling te beoordelen. de effectiviteit van vaste ultrafiltratie-membraan (10 kDa) als voorbehandeling voor het vertragen van organische/biofouling van SWRO voedingswater onderzocht. Dit onderzoek werd in laboratorium en met behulp van proef-machinerie verricht door verschillende MF/UF membranen en organische algenmaterie (AOM) te gebruiken die door *Chaetoceros affinis* als voedingsoplossing geproduceerd werden.

Het AOM-verwerpingsexperiment, dat uitgevoerd werd met MF/UF membranen, toonde aan dat vaste UF (10 kDa) een 3 tot 4 keer hogere biopolymer concentratie, TEP concentratie en MFI-UF (vergeleken met het hoge moleculaire gewichtsvermindering van MF/UF membranen) kunnen verminderen. De gemeten bacterial regrowth potential (BRP) van de vaste UF permeaat sample was ook ongeveer 2 tot 3 keer lager dan in permeaat van de hogere moleculaire gewichtsvermindering van het MF/UF membraan. Bovendien werd er geen opmerkelijk verschil gevonden in de bacterial regrowth potential (BRP) van vaste UF (10 kDa) en hoge moleculaire gewichtsvermindering (150 kDa UF) tijdens het uitvoeren van de experimenten. Biofouling experimenten, op schaal uitgevoerd met een permeaat bestaande uit vaste UF en conventionele UF (150 kDa), toonde geen substantiële ontwikkeling van drukverlies in het membrane Fouling Simulator (MFS) cellen aan gedurende het verkorte experiment van 15 dagen. Toch werd de biomassa, die zich in de MFS monitor ophoopte en gevoed werd met 10 kDa UF permeaat, gemeten op zo'n 860 pg-ATP/cm^2, wat 2 tot 5 keer lager dan wat gemeten werd in de MFS monitor; respectievelijk gevoed met 150 kDa UF permeaat en UF voeding. Met betrekking tot de hydraulische uitvoering toonde de vaste UF aan dat het een 1,5 keer hogere non-backwashable fouling rate ontwikkeling heeft vergeleken met 150 kDa UF. Dit kan worden verklaard door de lagere oppervlakte poreusheid van het 10 kDa UF membraan, wat resulteerde in een lagere backwashing en chemical enhanced backwashing (CEB) efficiëntie vergeleken met 150 kDa UF. De verbetering van de surface porosity van 10 kDa UF zou het verlagen van de ontwikkeling van de non-backwashable fouling ratio kunnen ondersteunen. Over het algemeen toonde de resultaten van de experimenten, die in het laboratorium op grote schaal uitgevoerd werden, aan dat het potentiele gebruik van vaste UF als pretreatment voor SWRO systemen tijdens algenbloei kan dienen, maar er is zeer zeker meer validatie voor deze uitkomst nodig door middel van grotere experimenten op een langer termijn.

Gedurende de laatste fase van de studie werd de rol van phosphate removal technology (PRT ™) onderzocht. Dit werd gecombineerd met vaste ultrafiltratie (10 kDa) die zou moeten leiden tot het elimineren of vertragen van de biofouling in SWRO membraansystemen. De experimenten die in het laboratorium uitgevoerd werden toonden aan dat de toepassing van PRT™ bijdraagt aan de hogere mate verwerping van biopolymer evenals opgelost phosphaat (fosfaat) uit SWRO voedingswater, vergeleken met de toepassing van ultrafiltratie op zichzelf. Verder ondersteunde de toepassing van

PRT™ gedeeltelijk de lagere bacterial regrowth potential (BRP) van z'n permeaat sample, losstaand van de kleine grote van het UF membraan gecombineerd met PRT™. De verdere toevoeging van 10 µgPO$_4$–P/L in permeaat van PRT™ toonde de hogere bacterial regrowth potential (BRP) aan, die de berperkende bacteriele groei door de verwijdering van fosfaat verklaart.

Uiteindelijk toonden de biofouling experimenten, uitgevoerd onder vergelijkbare omstandigheden in SWRO en bij gebruik van een membrane fouling simulator (MFS), geen toename aan in het voedingskanaal met een drukafname in MFS, gevoed met permeaat van vaste UF + PRT™, gedurende een periode van 21 dagen wanneer uitgevoerd bij een kruisstroom snelheid van 0.2 m/s. Bovendien werd er een drukafname van ongeveer 500 mbar ondervonden in MFS fed met een permeaat van vaste UF wanneer deze tijdens dezelfde tijdsperiode en onder vergelijkbare omstandigheden werd uitgevoerd, en uiteindelijk de rol van PRT™ in het vertragen van de aanwezigheid van biofouling illustreert. Het resultaat van membraan autopsie toonde aan dat de opeenstapeling van biomassa in MFS fed, met permeaat bestaande uit UF+ PRT™, onder de detectiegrens lagen. Dit, terwijl de gemeten ATP 6,000 pg-ATP/cm^2 was in de MFS fed met permeaat van vast UF op zichzelf. De hogere opeenstapelingen van biomassa in MFS fed, met permeaat van vaste UF, konden aan de doorgang van laag moleculair gewichts (LMW), organisch- en opgeloste fosfaat door UF toegewezen worden. De mogelijke bijdrage van LMW (getest door EDTA) toonde een lineaire hergroei ($R^2 = 0.65$) tussen de EDTA concentraties en de netto bacteriele hergroei aan. Over het algmeen demonstreerden de experimenten dat de verwijdering van fosfaat, door de toevoeging van PRT™ gecombineerd met UF (10 kDa), de groei van biomassa limiteert, en dus de aanwezigheid van biofouling in SWRO membrane vertraagt of zelfs elimineert. Bovendien zijn er meer experimenten nodig voor verdere verificatie.

Over het algemeen toonde de toevoeging van vaste UF (10 kDa) en Phosphate Removal Technology (PRT™) een hogere potentie naar de verwijdering van AOM en voedingsstoffen aan, en vertraagde de biofouling in SWRO systemen. Het onderzoek benadrukte ook het belang van het controleren van de waterkwaliteit tijdens de behandelingsprocessen door het gebruik van betere analytische methodes voor de ontwikkeling van membrane fouling beperkingsstrategieën. De vaststelling van MFI-UF$_{10kDa}$, biopolymer, TEP$_{10kDa}$ concentratie tijdens algenbloem, en de correlatie met de

MODIS satelliet-data, zal de bruikbare informatie over fouling potentie van zeewater op verschillende locaties genereren. Verder is het van belang dat de gedetailleerde onderzoeken, die verricht worden om de operationele stabiliteit van geteste voorbehandlingstechnologieën in SWRO systemen tijdens de algenbloem te testen, uitgevoerd worden op verschillende types verandering, coagulatie dosering etc. De verdere verbetering in materiele bezittingen moeten ook gezien worden als iets wat de kosten van de bediening kan verminderen en de stabiliteit van de hydraulische bediening kunnen verzekeren. Tot slot is het nog steeds van belang om methodes te ontwikkelen die lage koolstof- en fosfaatconcentraties in zeewater kunnen detecteren, wat weer bij kan dragen aan de ontwikkeling van betere membraan fouling preventie strategieën.

Abbreviations

AB	Alcian blue
AOC	Assimilable organic carbon
AOM	Algal organic matter
ASW	Artificial seawater
ATP	Adenosine triphosphate
BDOC	Biodegradable organic carbon
BRP	Bacterial regrowth potential
BW	Brackish water/Backwashing
BWRO	Brackish water reverse osmosis
CEB	Chemical enhanced backwashing
CF	Cartridge filter
CFU	Colony forming unit
CIP	Cleaning- in- place
DAF	Dissolved air flotation
DMF	Dual media filters
DOC	Dissolve organic carbon
EBCT	Empty bed contact time
ED	Electrodialysis
EDTA	Ethylene-di-amine-tetra-acetic acid
EPS	Extracellular polymeric substances
FCM	Flow cytometry
FEEM	Fluorescence excitation emission matrix
FESEM	Field emission scanning electron microscopy
FTIR	Fourier transform infrared spectroscopy
GMF	Granular media filtration
HPC	Heterotrophic plate counts
LCOCD	Liquid chromatography with organic carbon detection
LMV	Lowest measured value
MED	Multi-effect distillation
MF	Microfiltration
MFI-UF	Modified fouling index -ultrafiltration
MFS	Membrane fouling simulator
MSF	Multi-stage flash
MWCO	Molecular weight cut off
NF	Nano filtration
NRV	Non return valve
OCD	Organic carbon detector
OND	Organic nitrogen detector
PC	Polycarbonate
PES	Poly-ether-sulphone

PI	Propidium iodide
PMA	Propidium monoazide
PRT™	Phosphate removal technology
PVDF	Poly-vinylidene-di-fluoride
RC	Regenerated cellulose
RO	Reverse osmosis
SDI	Silt density index
SEM	Scanning electron microscope
SLMB	Schweizerische Lebensmittelbuch
SW	Seawater
SWRO	Seawater reverse osmosis
TCC	Total cell count
TDC	Total direct cell
TDS	Total dissolved solid
TEP	Transparent exopolymeric particles
TMP	Transmembrane pressure
TOC	Total organic carbon
UF	Ultra filtration
UV	Ultraviolet
XRD	X-ray diffraction

Publications and awards

Journal publications

1. **Dhakal, N.,** Salinas Rodriguez, S.G., Schippers, J.C. and Kennedy, M.D. (2014) *Induction time measurements in two brackish water reverse osmosis plants for calcium carbonate precipitation*, Desalination and Water Treatment, doi: 10.1080/19443994.2014.903870

2. **Dhakal, N.,** Salinas Rodriguez, S.G., Schippers, J.C. and Kennedy, M.D. (2014) *Perspectives and challenges for desalination in developing countries*, IDA journal of Desalination and Water Reuse, doi: 10.1179/2051645214Y.0000000015

3. Villacorte, L.O., Tabatabai, S.A.A., **Dhakal, N.,** Amy, G., Schippers, J.C. and Kennedy, M.D. (2014) *Algal blooms: An emerging threat to seawater reverse osmosis desalination*. Desalination and Water Treatment, doi: 10.1080/19443994.2014.940649

4. Shirleyana, **Dhakal N.,** Arevalo M. (2012). *Using GIS tools for identifying new family planning clinics based on social and accessibility factors, the case of central Java, Indonesia*. Computer Science Journal of Pelita Harapan University, ISSN 1412- 9523

Journal articles submitted/in preparation

1. **Dhakal, N.,** Salinas Rodriguez, S.G., Ouda, Alaa., Schippers, J.C. and Kennedy, M.D. (2017) Fouling of ultrafiltration membranes by organic matter generated by four marine algal species. Submitted to Journal of Membrane Science

2. **Dhakal, N.,** Salinas Rodriguez, S.G., Ampah, J., Schippers, J.C. and Kennedy, M.D. (2017) *Measuring bacterial regrowth potential (BRP) in seawater reverse osmosis using a natural bacterial consortium* and flow cytometry. Submitted to Desalination

3. **Dhakal, N.,** Salinas Rodriguez, S.G., Schippers, J.C. and Kennedy, M.D. (2017) *The role of tight ultrafiltration on reducing fouling potential of SWRO feed water during algal blooms*, In preparation to Desalination

4. **Dhakal, N.,** Salinas Rodriguez, S.G., Schippers, J.C. and Kennedy, M.D. (2017) *Phosphate removal in seawater reverse osmosis feed water: An option to control biofouling during algal blooms*, In preparation to Desalination

Presentations in international conferences

1. **Dhakal, N.,** Salinas Rodriguez, S.G., Schippers, J.C. and Kennedy, M.D. (2012) *Prediction of antiscalant dose for CaCO₃ scaling in BWRO systems*, presented in EDS Conference on Membrane in Drinking and Industrial Water Production, Leeuwarden, Netherlands, September 10 -12, 2012

2. **Dhakal, N.,** Salinas Rodriguez, S.G., Schippers, J.C. and Kennedy, M.D. (2012) *Perspective and challenges for desalination in developing countries: Where, when and how should desalination be implemented?* presented in IWA World Water Congress and Exhibition, Busan, South Korea, September 16 - 21, 2012

3. **Dhakal, N.,** Salinas Rodriguez, S.G., Schippers, J.C. and Kennedy, M.D. (2012) *Minimizing the consumption of chemicals by improving anti-scalant dose prediction for CaCO₃ scaling in BWRO systems*, presented in EDS Conference and Exhibition on Desalination for the environment: Clean Water and Energy, Barcelona, Spain, April 22 - 26, 2012

4. **Dhakal, N.,** Salinas Rodriguez, S.G., Schippers, J.C. and Kennedy, M.D. (2014) *Improving seawater reverse osmosis (SWRO) pretreatment to reduce biofouling potential during algal blooms*, presented in Euro Med Conference on Desalination for Clean Water and Energy, Palermo, Italy, May 10 - 14, 2014

5. **Dhakal, N.,** Nazeer, A., Villacorte, L.O., Salinas Rodriguez, S.G., Schippers, J.C. and Kennedy, M.D. (2014) *Role of UF pre-treatment in controlling the biological growth potential in SWRO*, presented in EDS Conference on Desalination for the Environment Clean Water and Energy, Limassol, Cyprus, May 11 - 15, 2014

6. **Dhakal, N.,** Villacorte, L.O., Salinas Rodriguez, S.G., Chauveau, S., Knops, F., Schippers, J.C. and Kennedy, M.D. (2015) *Algal blooms and advanced pre-treatment in seawater reverse osmosis*, presented in IDA World Congress: Desalination and Water Reuse, Renewable Water Resources to Meet Global Needs, San Diego, USA, September 28 - November 4, 2015

7. **Dhakal, N.,** Ouda, A., Salinas Rodriguez, S.G., Schippers, J.C. and Kennedy, M.D. (2016) *Fouling propensity of marine bloom forming algae in membrane based systems,* presented in EDS Conference on Desalination for Clean Water and Energy, Palermo, Italy, May 22 - 26, 2016

8. **Dhakal, N.,** Salinas Rodriguez, S.G., Schippers, J.C. and Kennedy, M.D. (2016) *Advanced pre-treatment to control biofouling in seawater reverse osmosis systems during algal blooms*, presented in EDS conference on Desalination for Clean Water and Energy, Palermo, Italy, May 22 - 26, 2016

9. **Dhakal, N.,** Salinas Rodriguez, S.G., Schippers, J.C. and Kennedy, M.D. (2016) *Innovative pre-treatment technologies for sustainable operation of Seawater Reverse Osmosis systems during algal blooms*, presented in the 13th IWA Leading Edge Conference on Water and Wastewater Technologies, Jerez de la Frontera, Spain, June 13 - 16, 2016

10. **Dhakal, N.,** Ouda, A., Salinas Rodriguez, S.G., Schippers, J.C. and Kennedy, M.D. (2017) *Fouling of ultrafiltration membrane during algal blooms*, presented in EDS Conference on Membrane in Drinking and Industrial Water Production, Leeuwarden, Netherlands, February 6 -8, 2017

Awards and fellowships

1. **Best oral presentation award** at UNESCO-IHE annual PhD Symposium, 2014
2. **MSc Fellowship** from Coca-Cola to pursue MSc degree at UNESCO-IHE, Delft (2009-2011)
3. **MSc Fellowship** from German Government (DAAD) to pursue MSc degree at University of Dortmund, Germany and University of the Philippine, Diliman (2007-2009)
4. **Best staff award** by the 13th District council of the Salyan District, Nepal (2005-2006)
5. **Fellowship** from Institute of Engineering, Nepal to purse BE in Civil Engineering (1998-2002)

Curriculum Vitae

Nirajan Dhakal

Oct. 2017 – present	Lecturer in Water Supply Engineering, Environmental Engineering and Water Technology Department, IHE Delft Institute for Water Education, the Netherlands
Dec. 2013 – Nov. 2017	Ph.D. research fellow at UNESCO-IHE, Wetsus European centre of excellence for sustainable water technology and TU Delft, the Netherlands
May 2011 – Nov. 2013	Researcher at UNESCO-IHE, the Netherlands
Oct. 2009 – Apr. 2011	Master of Science in Water Supply Engineering at UNESCO-IHE, the Netherlands
Sep. 2007 – Sep. 2009	Master of Science in Regional Development Planning and Management at University of Dortmund, Germany, and University of the Philippines
May 2006 – Aug. 2007	Water Supply and Sanitation Engineer at Action Contre La Faim (ACF, Nepal Mission)
Aug. 2003 – Jul. 2007	Civil Engineer at Government of Nepal, District Development Committee, Salyan, Nepal (For project funded by United Nations Capital Development Fund (UNCDF) & British Governments Department for International Development-DFID)
Dec. 2002 – Jul. 2003	Lecturer at Kathmandu Institute of Technology, Nepal
Sep. 1998 – Nov. 2002	Bachelor in Civil Engineering, Institute of Engineering, Pulchowk, Nepal
26 June 1979	Born in Shreenathkot-3, Gorkha, Nepal

This PhD research work was carried out at UNESCO-IHE with the financial support from Wetsus, European Centre of Excellence for Sustainable Water Technology (www.wetsus.nl). Wetsus is funded by the Dutch Ministry of Economic Affairs, the European Union Regional Development Fund, the Province of Fryslân, the City of Leeuwarden and the EZ/Kompas program of the "Samenwerkingsverband Noord-Nederland".

Netherlands Research School for the
Socio-Economic and Natural Sciences of the Environment

D I P L O M A

For specialised PhD training

The Netherlands Research School for the
Socio-Economic and Natural Sciences of the Environment
(SENSE) declares that

Nirajan Dhakal

born on 26 June 1979 in Gorkha, Nepal

has successfully fulfilled all requirements of the
Educational Programme of SENSE.

Delft, 30 November 2017

the Chairman of the SENSE board

Prof. dr. Huub Rijnaarts

the SENSE Director of Education

Dr. Ad van Dommelen

The SENSE Research School has been accredited by the Royal Netherlands Academy of Arts and Sciences (KNAW)

K O N I N K L I J K E   N E D E R L A N D S E
A K A D E M I E   V A N   W E T E N S C H A P P E N

The SENSE Research School declares that Mr Nirajan Dhakal has successfully fulfilled all requirements of the Educational PhD Programme of SENSE with a work load of 40 EC, including the following activities:

SENSE PhD Courses

o Environmental research in context (2014)
o Research in context activity: 'Organizing and reporting on a research campaign at Desalination Plant (Ras Al Khaimah, UAE - 18-27 September 2016)'

Other PhD and Advanced MSc Courses

o Design course on water treatment with X-Flow membrane products, Pentair X-FLow Academy (2015)
o Membrane technology in drinking and industrial water treatment, UNESCO-IHE (2013)

Management and Didactic Skills Training

o Supervising six MSc students (2011-2017)
o Teaching lab session on membrane filtration in the MSc course module 'Advanced Water Treatment and Reuse' UNESCO-IHE (2015-2017)
o Co-organising International Conference on Membranes in Drinking and Industrial Water production (MDIW), 6-8 February 2017, Leeuwarden, The Netherlands
o Co-organising UNESCO-IHE PhD symposium 'From Water Scarcity to Water Security - Achieving the Sustainable Development Goals', 3-4 October 2016, Delft, The Netherlands
o Moderating session 'Safe drinking water and sanitation' in UNESCO-IHE PhD symposium 'Urban Sustainability', 29-30 September 2014, Delft, The Netherlands
o Participating as a member in the PhD Association Board of UNESCO-IHE (2014-2016)

Selection of Oral Presentations

o *Fouling of ultrafiltration membranes during algal blooms.* International Conference on Membranes in Drinking and Industrial Water production (MDIW), 6-8 February 2017, Leeuwarden, The Netherlands
o *Fouling propensity of marine bloom forming algae in membrane based systems.* European Desalination Society (EDS) Conference on Desalination for the Environment: Clean Water and Energy, 22-26 May 2016 Palermo, Italy
o *Algal blooms and advanced pre-treatment in seawater reverse osmosis.* IDA World Congress 'Desalination and Water Reuse', 30 August - 4 September 2015, San Diego, USA
o *Role of UF pre-treatment in controlling the biological growth potential in SWRO.* European Desalination Society (EDS) Conference on Desalination for Environment: Clean Water and Energy, 11-15 May 2014, Limassol, Cyprus

SENSE Coordinator PhD Education

Dr. ing. Monique Gulickx

For Product Safety Concerns and Information please contact our EU
representative GPSR@taylorandfrancis.com Taylor & Francis Verlag GmbH,
Kaufingerstraße 24, 80331 München, Germany

Printed and bound by CPI Group (UK) Ltd, Croydon, CR0 4YY
01/05/2025
01858618-0002